# The Politics of Coal's Decline

*Research by the Energy and Environmental Programme is supported by generous contributions of finance and professional advice from the following organizations:*

*AEA Technology • Amerada Hess • Arthur D Little*
*Ashland Oil • British Coal • British Nuclear Fuels*
*British Petroleum • European Commission*
*Department of Trade and Industry • Eastern Electricity*
*Enterprise Oil • ENRON Europe • Exxon • LASMO*
*Mobil • National Grid • National Power • Nuclear Electric*
*Overseas Development Administration • PowerGen*
*Saudi Aramco • Shell • Statoil • St Clements Services*
*Texaco • Total • Tokyo Electric Power Company*

# The Politics of Coal's Decline

## The Industry in Western Europe

Mike Parker

THE ROYAL INSTITUTE OF INTERNATIONAL AFFAIRS
**Energy and Environmental Programme**

EARTHSCAN
Earthscan Publications Ltd, London

First published in Great Britain in 1994 by
Royal Institute of International Affairs, 10 St James's Square, London SW1Y 4LE and
Earthscan Publications Ltd, 120 Pentonville Road, London N1 9JN

Distributed in North America by
The Brookings Institution, 1775 Massachusetts Avenue NW,
Washington DC 20036-2188



A catalogue record for this book is available from the British Library.

Paperback: ISBN 1 85383 248 0

The Royal Institute of International Affairs is an independent body which promotes the rigorous study of international questions and does not express opinions of its own. The opinions expressed in this publication are the responsibility of the author.

Earthscan Publications Limited is an editorially independent subsidiary of Kogan Page Limited and publishes in association with the International Institute of Environment and Development and the World Wide Fund for Nature.

Printed and bound by Biddles Limited, Guildford and King's Lynn.
Cover by Visible Edge.
Illustration by Andy Lovel.

# Contents

Acknowledgements ........ viii
Preface ........ ix
Abbreviations ........ x
Summary and Conclusions ........ xi

**Introduction** ........ 1

**1. The Background: Hard Coal in Western Europe** ........ 3
1.1 Overall trends in production and demand since 1973 ........ 3
1.2 Recent trends in the European Union ........ 5
1.3 Projected trends in production and demand to 2005 ........ 7
1.4 Trends in individual countries ........ 9
1.5 The issue of coal-fired generating plant ........ 18
1.6 Emergence of environmental issues ........ 19
1.7 Issues for the future ........ 23

**2. The Problem of High-Cost Production** ........ 27
2.1 The extent of uncompetitiveness ........ 27
2.2 The subsidy regime to 1992 ........ 29
2.3 The new Decision on state aids: the problems of agreement ........ 36
2.4 The new Decision on state aids: terms and conditions ........ 42

**3. The Future of Protection in Europe** ........ 47
3.1 Introduction: the EU-12 context ........ 47
3.2 The future of protection: the United Kingdom ........ 48
3.3 The future of protection: Germany ........ 50
3.4 The future of protection: Spain ........ 54
3.5 The future of protection: France ........ 55

**4. The Environment and the Politics of Coal** .......... 57
4.1 'Acid rain' .......... 57
4.2 Clean coal technology .......... 61
4.3 Climate change and carbon dioxide .......... 64

**5. Policy Synthesis: Prospects and the Politics of Coal in Europe** .......... 71

**Tables**

1.1 Western European hard coal balance, 1973 and 1993 .......... 4
1.2 Western European hard coal balance by country, 1993 .......... 5
1.3 Changes in hard coal production and net imports between 1973 and 1993 .......... 7
1.4 EU-12 hard coal balance .......... 8
1.5 EU-12 hard coal market, 1993 and 1994 .......... 9
1.6 EU-12 hard coal production, 1992, 1993 and 1994 .......... 9
1.7 EU-12 net coal imports, 1993, by country .......... 10
1.8 Future hard coal market .......... 11
1.9 Future hard coal production .......... 12
1.10 Projected changes in hard coal production and net imports between 1993 and 2005 .......... 13
1.11 Projected change in import dependence, 1993–2005 .......... 14
1.12 Existing generating plant using hard coal .......... 19
1.13 New generating plant using hard coal .......... 20
1.14 Sulphur reduction targets in the Large Combustion Plant Directive .......... 21
1.15 Sulphur reduction targets in the UNECE Second Sulphur Protocol .......... 22
1.16 Relative emissions of $SO_2$ and $NO_x$ from different generating technologies .......... 23
1.17 $CO_2$ emission reduction targets in the EU-12 .......... 24
1.18 Relative emissions of $CO_2$ from different generating technologies .......... 25
2.1 Trend in producer subsidy equivalents for hard coal in Europe .......... 28
2.2 Trend of imported coal prices .......... 29

2.3 Volume, cost and productivity of coal production in Europe ........... 30
2.4 Productivity and labour changes in European coal production, 1992–1993 ........... 31
2.5 State aids for coal production and adjustment in Europe, 1992 ...... 33
2.6 Variation in revenue, PSE and state aids in Europe ........... 35
3.1 The German coal industry: summary 1992–1993 ........... 53
4.1 $CO_2$ emissions and hard coal intensity, 1990 ........... 66
4.2 Energy-related $CO_2$ emissions per unit of GDP ........... 67
5.1 Coal's share of European primary energy by 2015/2020 under different scenarios ........... 74

**Figure**

1.1 European hard coal production ........... 6

# Acknowledgements

The author wishes to thank those who have provided constructive comment in the preparation of this study, in particular Anthony Baker, Matthias Bruch, Christian Cleutinx, Dave Feickart, Matthias Hessling, Walter Jensen and Walt Patterson.

Thanks are also due to Michael Grubb and the staff of the Energy and Environmental Programme at Chatham House. However, any views expressed and any errors, whether of fact or judgement, remain the responsibility of the author.

*November 1994* Mike Parker

# Preface

The industrial revolution was powered by coal. So was Europe's recovery from the Second World War. Yet across Europe, ever since the early 1960s, coal has been facing the pressure of an industry being progressively weakened by alternative sources of industrial and power-station fuel: first oil, then nuclear, and increasingly now both gas and lower-cost coal imports. As the perceived threat of high oil prices has receded, environmental costs have risen and the desire to open up energy markets has spread across Europe, the coal industry's future looks increasingly bleak.

That is one of the messages to emerge from Mike Parker's study, which takes a clear and objective, if painful, look at the European prospects in the aftermath of the UK 'coal crisis'. The other main message to emerge is the extraordinary impact of politics on the rate and brutality of the process of contraction. He sets out graphically the stark contrast between policies in the UK and in Germany, showing how political support for the coal industry in the two countries stands in direct opposition to its relative economic strength.

As the former Director of Economics at British Coal, Mike Parker brings to bear a rare degree of knowledge and personal experience concerning the economic and political issues involved. The study builds upon his earlier report *The UK 'Coal Crisis': Origins and Resolution*, and completes a portfolio of studies on the fossil fuel and electricity industries in Europe published by our Programme over the past two years. With European energy policy under review in the run-up to the 1996 intergovernmental conference, it is a valuable contribution to the debate and we are most grateful to Mike Parker for completing the work so promptly and efficiently.

*November 1994*

Dr Michael Grubb
Head, Energy and Environmental Programme

# Abbreviations

| | |
|---|---|
| BATNEEC | Best available technique not involving excessive cost |
| BC | British Coal |
| CCGT | Combined-cycle gas turbine |
| CEPCEO | Western European Coal Producers' Association |
| CIF | (Price including) Cost, insurance and freight |
| ECSC | European Coal and Steel Community |
| ECU | European Currency Unit |
| EU-12 | European Union |
| FGD | Flue Gas Desulphurization |
| GJ | Gigajoule |
| GW | Gigawatt |
| HMIP | Her Majesty's Inspectorate of Pollution |
| ICR | International Coal Report |
| IEA | International Energy Agency |
| HUNOSA | Hulleros del Norte (Spain) |
| LCPD | Large Combustion Plant Directive |
| mtce | Million tonnes coal equivalent (29.3 GJ/tonne) |
| NUM | National Union of Mineworkers |
| PSE | Producer subsidy equivalent |
| tce | Tonnes coal equivalent (29.3 GJ/tonne) |
| UNECE | United Nations Economic Commission for Europe |

# Summary and Conclusions

The study considers past and future policies for hard coal production in the European Union against the background of trends in the market for hard coal in western Europe. By 1993, coal's share of the primary energy market in the region had fallen to 13%, with EU-12 coal production, which had halved over the previous 20 years, providing only 7%. Imports from outside the EU-12 made up nearly half of the total hard coal supply. 'Conventional' projections show the coal market broadly stable to 2005, but with EU-12 production falling by a further 40% and import dependence rising to two-thirds. Such projections almost certainly overstate total coal demand and the decline in EU-12 production.

Now that the OPEC threat appears to have receded, and gas supplies appear abundant, the political influences in favour of coal use (as distinct from protection of indigenous production) have virtually disappeared, and have been replaced by environmental concerns. In the long term, the market for coal is likely to depend upon the commercial availability of new, advanced 'clean coal' electricity-generating plant to replace the ageing stock of conventional plant, but there is no sign that the present R&D programme will be sufficient to allow this to happen. In the meantime, while there is no evidence that any EC/EU environmental regulation has been adopted which would limit against its government's wishes any member state's ability to use coal, on the other hand it remains highly unlikely that the Commission would seek to intervene in the market to protect coal from its environmental disadvantages, or that any member state would seek to do so, except in support of a policy linked to a political imperative to protect indigenous production.

A centrally directed Community policy to protect indigenous coal production has never been a serious possibility, because of the lack of common interest. Over 80% of EU-12 hard coal production is in Germany and the UK, with the balance in Spain and France. The minority of member states which produce coal have had decreasing influence as the Community has been enlarged, with most member states concerned to retain access to low-cost coal imports rather than to support high-cost EU-12 production. Instead, the subsidiarity principle and domestic politics have governed the policies of the coal-producing member states towards their indigenous coal industries (including the impact of environmental regulations), in circumstances where the policies of individual member states have had no practical effect on the policies of the others.

This has led to widely differing results, and will continue to do so. In the UK there has been rapid contraction and privatization; in Germany, massive subsidies and slow adaptation by consensus; in Spain, selective restructuring and subsidies; and in France, slow phasing out. The 'Anglo-German anomaly' is particularly striking: the broad aim of German policy is to sustain by subsidy a deep-mine production twice that likely to survive in the UK, even though German costs are likely to remain some three times the UK level. This marked divergence of policy can be explained only by wide differences in domestic politics. In the UK, the Conservative government's political aim irreversibly to break the power of the National Union of Mineworkers has had no parallel in Germany; and in Germany, the checks and balances of the federal structure have sustained the influence of the coal-producing regions, whereas in the UK there are no such regional checks of sufficient influence to mitigate the rigour of central government policy.

If there were another serious nuclear accident, if alarm about the security of Russian and Algerian gas supplies were to grow, if global warming were no longer thought important, then the politics of coal could be changed sufficiently to stimulate investment in new coal-fired generation plant, thereby sustaining coal consumption and slowing the contraction of the remaining EU-12 coal production. But the most important issue is likely to continue to be the politically acceptable phasing of decline: a process governed by domestic politics operating within the subsidiarity principle. This is a political, not an economic, concept and, in the case of coal, has led to outcomes so different in the UK and Germany as to strain the bounds of credibility - until we remember the politics.

# Introduction

At the end of *The UK 'Coal Crisis': Origins and Resolution* I wrote:

> It is difficult to avoid the conclusion that the 'coal crisis' of 1992/3 was not about fundamental strategic direction, but about politically acceptable phasing at the beginning of the last chapter in the story of a once great industry, which until recently was widely expected to make a continuing large contribution to future UK energy needs. And what has happened to the UK coal industry will cast its shadow over the rest of the coal industry in the European Community.[1]

This study deals with the past and future policies for EU-12 hard coal production, against the background of trends in the market for hard coal in western Europe.[2] The analysis shows that the story will be one of contraction: but the extent and manner of the decline in western European coal production will be determined as much by political as economic factors. Moreover, the coal issue also serves to illustrate two important general aspects of policy, namely how the interaction of environmental and energy policies works in practice, and the extent to which the affairs of the European Union are, or are not, resolved by collective action, when set against the particular interests of individual member states.

---

[1] Mike Parker, *The UK 'Coal Crisis': Origins and Resolution*, London, RIIA, 1993, p. 45.

[2] This study does not deal directly with the brown coal/lignite industry, which is distinct from the hard coal industry in terms of its technology, economics and, to a considerable extent, environmental impacts. 'Western Europe' refers to OECD Europe excluding Turkey.

# Chapter 1

## The Background: Hard Coal in Western Europe

### 1.1 Overall trends in production and demand since 1973

This section presents the basic statistics of hard coal production and demand in western Europe since 1973, and examines the trends and pressures upon coal across the EU-12 and separately in the member states. Much of the analysis is based on statistics published by the International Energy Agency (IEA) in *Coal Information* (1994 edition), and is presented in terms of million tonnes 'coal equivalent' (mtce), using a standard coal quality of 29.3 GJ/tonne.[3]

The *market for hard coal* may be defined as production plus net imports. It may differ slightly from consumption due to stock changes. Table 1.1 summarizes these data for 1993, compared with the situation 20 years previously. Over the two decades since the first OPEC oil price 'shock', the western European market for hard coal declined (with much of this fall occurring in 1993), with a large reduction in indigenous production partly offset by an increase in coal imports from outside the EU-12. Moreover, over this period the total primary energy demand in western Europe showed an increase of 254 mtce (15%); so that hard coal's share of total primary energy supply declined from 17% in 1973 to 13% in 1993. Hard coal *production* provided only 7% of western European primary energy in 1993.

However, because of the nature of the western European coal market, aggregates are not very useful; Table 1.2 therefore shows production, net imports and the total market for the main individual countries. Virtually all the western European hard coal output comes from four EU-12 countries, and in 1993 86% of the total output was in the UK and Germany. Furthermore,

---

[3] This is close to the calorific value of prime quality coking coal, but most coal, particularly steam coal, has significantly less GJ/tonne, so that most coal statistics in unadjusted 'tonnes' will show higher volumes. The use of mtce ensures comparability.

**Table 1.1 Western European hard coal balance, 1973 and 1993 (mtce)**

| | 1973 | | | 1993[a] | | |
|---|---|---|---|---|---|---|
| | Production | Net imports | Total market | Production | Net imports | Total market |
| EU-12 member states | 252 | 39 | 291 | 135 | 103 | 238 |
| Other OECD Europe (excl. Turkey) | neg. | 7 | 7 | neg. | 13 | 13 |
| Total | 252 | 46 | 298 | 135 | 116 | 251 |

[a]partly estimated.
Source: IEA data.

the four 'coal producer' countries accounted for 73% of the total western European hard coal market. The latter figure, however, is not much bigger than the share of these countries in total European energy demand. In aggregate hard coal provided 15% of primary energy demand in these countries in 1993: little more than the western European average of 14%. Using the IEA data mentioned above, a graphical representation of the situation is given in Figure 1.1.

The present configuration of the western European hard coal market can be more readily understood by identifying the main changes that took place between 1973 and 1993. From Table 1.3 it can be seen that:

- In all countries where there has been a net increase in the market for coal, this has been based exclusively on imports from outside the EU-12 (except for the case of Spain, but even here most of the increase in the coal market has been attributable to imports).
- In those countries which were the major coal producers in 1973 – the UK, Germany and France – the fall in production exceeded the increased use of imports, so that there was a significant fall in the market for hard coal in these countries.

**Table 1.2 Western European hard coal balance by country, 1993[a] (mtce)**

| | Production | Net imports | Total market |
|---|---|---|---|
| UK | 56 | 14 | 70 |
| Germany | 60 | 10 | 70 |
| Spain | 10 | 11 | 21 |
| France | 8 | 14 | 22 |
| Total 'coal producers' | 134 | 49 | 183 |
| Italy | | 14 | 14 |
| Belgium | | 12 | 12 |
| Netherlands | | 11 | 11 |
| Denmark | | 9 | 9 |
| Other EU-12 | | 9 | 9 |
| Other OECD Europe | | 13 | 13 |
| Total 'non-producers' | | 68 | 68 |
| *Total* | 135[b] | 116 | 251 |

[a]Partly estimated.
[b]Totals include small amount of production from countries not normally considered 'coal producers'.
Source: IEA.

## 1.2 Recent trends in the European Union

This section looks in more detail at recent trends in the EU-12, using data drawn together by the European Commission on the market in different member states. The subsequent section will then summarize the important features of production and the overall market, and the main factors that have influenced these, on a country-by-country basis.

So far as the EU-12 is concerned, a Commission Report published at the end of 1993 showed that there has been a downturn in the total coal market, which is expected to continue into 1994, with EU-12 production being more seriously affected than imports from third countries (see Table 1.4).[4] In large measure, the Commission attributes this downturn to the effects of economic recession in Europe, and the consequential downturn in electricity generation and steel production. However, the Commission also notes that the large

[4] European Commission, *The Market for Solid Fuels in the Community in 1993 and the Outlook for 1994* (ENG.DOC), Brussels, 29 November 1993.

**Figure 1.1 European hard coal production (million tonnes)**

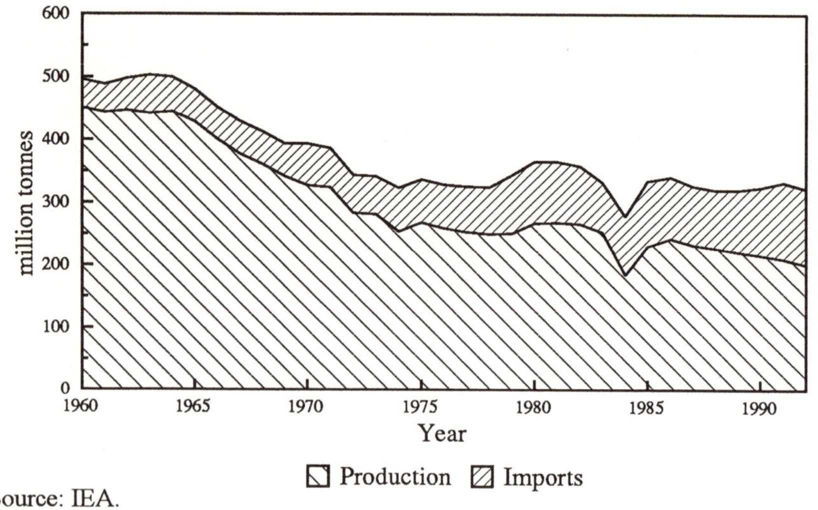


Source: IEA.

structural changes taking place in the UK coal market, in both supply and demand, are having a significant effect on the total EU-12 figures. The structure of the *market* for hard coal for 1993 and 1994 is set out in Table 1.5. It can be seen that, in the Commission's forecast for 1994, the reduction in the total EU-12 market for coal can be explained almost entirely by the reduction in the UK power station coal market. For the rest, there is something like stability.

On *production*, the Commission's figures are summarized in Table 1.6. Here again, the reduction in output is dominated by the UK (particularly as the Commission has probably underestimated the rundown in 1994), but the reduction in Germany has also been significant, continuing past trends.

The current degree of import dependence of individual EU-12 countries in 1993 varied enormously (see Table 1.7). Of the two biggest producers, Germany and the UK, less than 20% of total coal consumption was from imports, and though the UK was still the biggest absolute importer, Spain imported about 40% of its coal, and France about 60%. All the other OECD countries in western Europe are virtually 100% dependent on imports for their small levels of hard coal consumption (excluding the special case of Spitzbergen), and virtually all imports were from outside the EU-12 sources;

**Table 1.3 Changes in hard coal production and net imports between 1973 and 1993 (mtce)**

| | Production | Net imports | Total market |
|---|---|---|---|
| UK | –52 | +15 | –37 |
| Germany | –40 | +9 | –31 |
| France | –16 | +2 | –14 |
| Spain | +2 | +9 | +11 |
| Belgium | –9 | +5 | –4 |
| Netherlands | –2 | +9 | + 7 |
| Sub total | –117 | +49 | –68 |
| Denmark | — | +6 | + 6 |
| Italy | — | +2 | + 2 |
| Portugal | — | +4 | + 4 |
| Ireland | — | +3 | + 3 |
| Greece | — | +1 | + 1 |
| Sub total | — | +16 | +16 |
| Total EU-12 | –117 | +65 | –52 |
| Finland | — | +3 | + 3 |
| Sweden | — | +2 | + 2 |
| Others | — | — | — |
| Total western Europe | –117 | +70 | –47 |

Source: IEA.

intra-Community trade constituted less than 5 million tonnes, or less than 2% of the total western European coal market.

## 1.3 Projected trends in production and demand to 2005

In order to consider how far past trends will continue into the future, a 'conventional' view of future trends on the hard coal market up to the year 2005 is set out in Table 1.8. On these projections, whereas in 1993 some three-quarters of the western European coal market was in the four 'coal producer' countries, by 2005 this proportion would fall to about two-thirds –

**Table 1.4 EU-12 hard coal balance (million tonnes)[a]**

| | 1992 Actual | 1993 Estimate | 1994 Forecast |
|---|---|---|---|
| EU-12 production[b] | 190 | 164 | 145 |
| Third-country imports | 133 | 115 | 113 |
| Total | 323 | 279 | 258 |
| *Sales* | | | |
| Power stations[c] | 211 | 181 | 168 |
| Coke ovens | 59 | 53 | 51 |
| Other markets | 42 | 40 | 40 |
| Total | 312 | 274 | 259 |

[a] The Commission data are reported in physical tonnes, rather than mtce. Since the calorific value of most coal in the EU is well below the official 'coal eqivalent' of 29.3 GJ/t, the tonnage figures for the EU here cannot be compared with the data in the previous section, and indeed frequently exceed the total mtce for western Europe.
[b] Including tip coal.
[c] Including industrial and pithead power stations.
Source: European Commission.

but would still constitute the dominant share of the market. Overall, however, in view of the substantial fall that has already taken place in the coal market, particularly in the UK, these 'conventional' projections now appear optimistic – indeed, the European Commission's latest view is that the consumption of coal could remain stable or decline.[5]

The latest IEA projections of hard coal *production* are shown in Table 1.9. Basically these reflect 'unchanged policies'. If we take the *bottom* of the range of the coal market set out in Table 1.8 and the production projections in Table 1.9 we can see the changes implied between 1993 and 2005 in these 'conventional' projections. These are shown in Table 1.10. As with the trends over the period 1973-93, the projections show a continuing decline in indigenous production and an increase in imports.

The projected trends imply a further substantial increase in import dependence in the total western European hard coal market, as set out in Table 1.11.

[5] European Commission, *Preparation Document for Green Paper on New Guidelines in Energy Policy*, 22 July 1994.

**Table 1.5 EU-12 hard coal market 1993 and 1994 (million tonnes)**

| | 1993 (estimated) | | | | 1994 (forecast) | | | |
|---|---|---|---|---|---|---|---|---|
| | Power stations[a] | Coke ovens | Other | Total | Power stations[a] | Coke ovens | Other | Total |
| UK | 61.5 | 8.5 | 12.2 | 82.2 | 46.5 | 8.5 | 12.0 | 67.0 |
| Germany | 52.6 | 15.5 | 6.3 | 74.4 | 52.2 | 13.9 | 6.6 | 72.7 |
| France | 6.4 | 8.1 | 7.7 | 22.2 | 7.7 | 7.6 | 7.5 | 22.8 |
| Spain | 24.5 | 4.6 | 2.3 | 31.4 | 24.5 | 4.4 | 2.9 | 31.8 |
| Sub-total | 145.0 | 36.7 | 28.5 | 210.2 | 130.9 | 34.4 | 29.0 | 194. |
| Belgium | 5.9 | 5.3 | 2.1 | 13.3 | 5.2 | 5.2 | 2.4 | 12.8 |
| Netherlands | 7.7 | 4.1 | 1.2 | 13.0 | 9.4 | 4.0 | 0.8 | 14.2 |
| Denmark | 10.0 | — | 0.8 | 10.8 | 10.5 | — | 0.7 | 11.2 |
| Italy | 6.3 | 6.9 | 3.0 | 16.2 | 6.3 | 6.9 | 3.0 | 16.2 |
| Portugal | 3.9 | 0.3 | 0.8 | 5.0 | 3.3 | 0.4 | 0.9 | 4.6 |
| Luxembourg | — | — | 0.3 | 0.3 | — | — | 0.2 | 0.2 |
| Ireland | 1.9 | — | 1.1 | 3.0 | 1.9 | — | 1.1 | 3.0 |
| Greece | 0.6 | — | 1.6 | 2.2 | 0.6 | — | 1.6 | 2.2 |
| Total EU-12 | 181.3 | 53.3 | 39.4 | 274.0 | 168.1 | 50.9 | 39.7 | 258.7 |

[a]Including industrial and pithead power stations.
Source: European Commission.

**Table 1.6 EU-12 hard coal production 1992, 1993 and 1994 (million tonnes)[a]**

| | 1992 | 1993 (estimated) | 1994 (forecast) |
|---|---|---|---|
| UK | 83.9 | 68.0 | 55.0 |
| Germany | 72.2 | 64.3 | 60.4 |
| Spain | 18.6 | 18.9 | 18.5 |
| France | 9.5 | 8.8 | 8.2 |
| Total | 184.2 | 160.0 | 142.1 |

[a]Excludes tip coal and very small tonnages in Portugal and Italy.
Source: European Commission.

## 1.4 Trends in individual countries

Both past trends and 'conventional' projections based on unchanged policies show marked differences between countries.

**Table 1.7 EU-12 net coal imports in 1993, by country**

| | Million tonnes | As % of total inland sales |
|---|---|---|
| UK | 15.5 | 18.9 |
| Germany | 10.2 | 13.7 |
| France | 13.7 | 61.8 |
| Spain | 12.5 | 39.8 |
| Belgium | 12.4 | 93.3 |
| Netherlands | 13.0 | 100.0 |
| Denmark | 10.8 | 100.0 |
| Italy | 14.9 | 100.0[a] |
| Portugal | 4.8 | 100.0[a] |
| Luxembourg | 0.3 | 100.0 |
| Ireland | 3.0 | 100.0 |
| Greece | 2.2 | 100.0 |

[a] Allowing for stock movements.
Source: European Commission.

*United Kingdom*

Between 1973 and 1993, the market for hard coal in the UK declined by about 35%, with coal supply to power stations showing little net change (but with both rises and falls within the period), while the coking, industrial and domestic markets showed substantial decline. In 1973 imports and exports were nearly in balance, but by 1993 there was a net import of 15 million tonnes. Over the period, deep-mined output more than halved, offset only to a limited extent by an increase in opencast output. The conventional projections, in Table 1.9, show a further halving of UK hard coal output by 2000. In fact this level will be reached very shortly, while the projected level of UK coal demand in 2000, shown in Table 1.8, has already been reached.

I have set out the genesis of this collapse in *The UK 'Coal Crisis': Origins and Resolution*.[6] A summary will suffice here. The Conservative government's policy objectives for the coal industry emerged as early as 1979-80, and had

[6] Parker *The UK 'Coal Crisis'*.

**Table 1.8 Future hard coal market (mtce)**

| | 1993[a] | 2000 | 2005 |
|---|---|---|---|
| UK | 70 | 59-76 | 60-70 |
| Germany | 70 | 73-95 | 72-91 |
| France | 21 | 26-36 | 25-31 |
| Spain | 22 | 24-28 | 24-30 |
| Sub-total 'coal producers' | 183 | 182-235 | 181-222 |
| Italy | 14 | 19-21 | 18-20 |
| Belgium | 12 | 9-12 | 9-12 |
| Netherlands | 11 | 10-16 | 10-16 |
| Denmark | 9 | 7-8 | 6-7 |
| Greece | 1 | 2-4 | 3-4 |
| Ireland | 3 | 2-3 | 2-3 |
| Portugal | 5 | 5-6 | 5-6 |
| Luxembourg | neg. | neg. | neg. |
| Sub total others | 55 | 54-70 | 53-68 |
| Total EU-12 | 238 | 236-305 | 234-290 |
| Finland | 6 | 8-11 | 8-11 |
| Sweden | 3 | 3-5 | 3-5 |
| Austria | 3 | 4-5 | 4-5 |
| Norway/Switzerland | 1 | 1-1 | 1-1 |
| Total | 251 | 252-327 | 250-312 |

[a]Partly estimated.
Source: author's projections, based on IEA Coal Information, 1994 edn., Martin C. Daniel, European Coal Prospects to 2010, Paris: IEA Coal Research, 1993.

a strong political element: to break the power of the National Union of Mineworkers (NUM), and to expose the industry to market forces as a prelude to its 'ultimate privatization'. These views were held strongly by Mrs Thatcher, who said in her memoirs that 'by the 1970s the coal mining industry had come to symbolize everything that was wrong with Britain'.[7] These elements of policy reinforced each other, leading to measures to reduce the size of the industry and its workforce as rapidly as possible.

Moreover, the political objectives went 'with the grain' of events, particularly after the defeat of the NUM strike in 1984-5 (which Mrs Thatcher saw as a

[7] Margaret Thatcher *The Downing Street Years*, London, HarperCollins, 1993, p. 340.

**Table 1.9 Future hard coal production (mtce)**

| | 1993[a] | 2000 | 2005 |
|---|---|---|---|
| UK | 56 | 29 | 21 |
| Germany | 60 | 50 | 50 |
| Spain | 10 | 11 | 10[b] |
| France | 8 | 7 | 5 |
| Belgium | 1 | — | — |
| Total | 135 | 97 | 86 |

[a]Partly estimated
[b]Assumed (data not included in IEA Report)
Source: IEA, Coal Information, 1994.

great symbolic victory which established 'that Britain could not be made ungovernable by the Fascist Left') and the relaxed international energy market following the collapse of oil prices in 1986.[8] Notwithstanding large increases in productivity, British Coal (BC) output remained uncompetitive with imports. Increasing environmental concerns about coal, which the government was, if anything, anxious to encourage, and the effects of electricity privatization (notably the large programme of new gas-fired generating plant) made it inevitable that BC would face further substantial contraction when its three-year contracts with the electricity supply industry expired in March 1993. In spite of the public and political outcry in October 1992 over the programme of colliery closures likely to arise, the outcome was a fall in the coal contract tonnages from 65 million tonnes in 1992/3 to 40 million tonnes in 1993/4 and 30 million tonnes in 1994/5, with the contraction in demand being reflected in reductions in deep-mined output.

It is worth examining further the major factor in the decline of UK coal demand, and hence production: namely the programme of combined-cycle gas turbine (CCGT) generating plant (popularly known as the 'dash for gas'). In the lead-up to electricity privatization, it was widely expected that the main challenge to BC from March 1993 would come from greatly increased coal imports by the main electricity generators. Figures in the region of 30 million tonnes were often quoted. This has not materialized: instead there has

[8] Ibid p. 378.

**Table 1.10 Projected changes in hard coal production and net imports between 1993 and 2005 (mtce)**

| | Production | Net imports | Total market |
|---|---|---|---|
| UK | –35 | +25 | –10 |
| Germany | –10 | +12 | +2 |
| Spain | — | +2 | +2 |
| France | –3 | +7 | +4 |
| Belgium | –1 | –2 | –3 |
| Sub total | –49 | +44 | –5 |
| Italy | — | +4 | +4 |
| Netherlands | — | –1 | –1 |
| Denmark | — | –3 | –3 |
| Greece | — | +2 | +2 |
| Ireland | — | –1 | –1 |
| Portugal | — | — | — |
| Total EU-12 | –49 | +45 | –4 |
| Finland | — | +2 | +2 |
| Sweden | | — | — |
| Austria | | +1 | +1 |
| Norway | | — | — |
| Total western Europe | –49 | +48 | –1 |

Source: Author.

been the 'dash for gas' which was in the main the result of regulatory policy. The electricity regulator was required, as an important part of his terms of reference, to promote 'competition' in generation. This was seen as requiring the encouragement of new entrants into generation to reduce the dominant share of National Power and PowerGen (itself the result of the failed attempt to incorporate nuclear power within the privatization process). Yet, as the generating capacity and fuel use of National Power and PowerGen were at the time overwhelmingly coal-based, any attempt to reduce their market share would necessarily also reduce their use of coal.

As no divestment of existing plant to competitors was then envisaged, the introduction of new entrants into generation meant that new generating plant not owned by the two major generators would have to be built, and the most cost-effective new plants were CCGTs. Both the regional electricity companies and oil companies supplying gas had other incentives to enter into this process,

**Table 1.11 Projected change in import dependence, 1993-2005 (mtce)**

| | 1993[a] | 2005 |
|---|---|---|
| Coal market | 251 | 250 |
| Production | 135 | 86 |
| Net imports | 116 | 164 |
| Import dependence | 46% | 66% |

[a]Partly estimated.
[b]Projected.
Source: Author.

using long-term contracts, and National Power and PowerGen also built CCGTs to defend their market share. In this way a large number of CCGTs, planned to operate at high load factors, were effectively committed which by the mid-1990s would take virtually half of BC's market. Yet in the majority of cases, the total costs of CCGTs (including return on capital) were higher than the avoidable costs of existing coal-fired plant, and the scale of the 'dash for gas' went a long way beyond what was required to facilitate achievement of environmental regulations on $SO_2$ emissions. It might be considered perverse to argue that the regulatory regime which gave rise to the 'dash for gas' was designed by government so as to halve the market for coal as rapidly as possible: yet this was the effect which the government made no attempt to mitigate, and indeed sought to defend in the White Paper *The Prospects for Coal.*[9]

*Germany*

Between 1973 and 1993 the market for hard coal in Germany declined by about 30%, with significant declines in the iron and steel, other industrial and residential markets. However, hard coal supply to power stations remained broadly stable, while most of the increase in German power requirements came from an expansion of nuclear power. Imports of hard coal have remained

[9] *The Prospects for Coal: Conclusions of the Government's Coal Review*, Cmnd 2235, London, HMSO, 1993.

stable, but exports of coking coal have declined. Hard coal production has shown a steady decline, but the industry has been heavily protected, and the contraction of output has been a carefully controlled process. The main preoccupation of government policy on hard coal has been to provide the mechanisms of subsidy and other forms of support sufficient to allow the industry to contract gradually at a rate consistent with agreement between government, industry and unions. The politics of coal have been greatly influenced by the federal constitution of Germany, which has enabled the coal industry to exercise a degree of influence which would not have been possible if the mines had been spread evenly across all *Länder*. For the future, the projections set out above imply that although there will be a steady reduction in hard coal production, demand will be maintained, leading to a significant increase in steam coal imports. This view is based on broad stability of nuclear output, and only a modest increase in the use of gas in power stations.

*France*

Between 1973 and 1993, the market for hard coal in France declined by 40%, with significant declines in the iron and steel and residential markets. However, hard coal supply to power stations remained broadly stable, but at a time when the output of French electricity more than doubled – reflecting the enormous increase in nuclear power – so that coal's share of electricity generation more than halved to less than 10%. Imports of hard coal have increased, albeit not dramatically. Hard coal production, which has been heavily protected, has fallen by nearly two-thirds. French government energy policy has been dominated by the nuclear power programme, leaving coal in a subsidiary role in power generation, and with only a third of the French coal market being met by indigenous production. For the coal industry, the main issue has been management of decline. Little change in the direction of policy is expected over the next decade, with indigenous production largely phased out by 2005 or shortly thereafter.

*Belgium*

Between 1973 and 1993, the market for hard coal declined by 25%, mainly in the iron and steel and residential markets. Hard coal supply to power stations showed broad stability, but at a time when there was a large increase in the use of nuclear power, both to provide for growth in electricity demand and to substitute for fuel oil. Imports of hard coal doubled, but by 1993 Belgian hard coal production (which had been heavily protected) had virtually ceased. The main coal policy issue for the Belgian government over the years was how to close down the coal industry without excessive social cost or disruption. With this now achieved, coal will play a declining subsidiary role in electricity generation.

*Spain*

Whereas most countries which were significant coal producers in 1973 (the UK, Germany, France and Belgium) saw a fall in their total market for coal, in Spain the market for coal doubled over the period 1973-93. While non-power markets declined slightly in total, there was a large increase in supplies of hard coal to power stations. Spanish electricity demand broadly doubled over this period. The increase in the use of coal complemented expansion in the use of brown coal and nuclear power, and a substantial reduction in the use of fuel oil. However, most of the increase in coal use has come from imported steam coal (mainly for power stations), although indigenous production has shown a small increase. The government's policy priorities have been to use coal as an alternative to nuclear power, when difficulties arose with the nuclear programme; and to achieve a manageable reconstruction of the coal industry, which contains elements that have been heavily subsidized. The above projections imply no significant change in these policies.

*Denmark*

Denmark has no coal production and imports all its requirements, with nearly all its coal market in power generation. The Danish electricity system is overwhelmingly dependent on coal, consumption of which increased four-fold since 1973, reflecting growth in electricity demand, substitution for fuel oil and the absence of nuclear power. As a result of past policies, Denmark is the most coal-dependent (and coal import-dependent) country in Western Europe. However, by 2005 Denmark plans to reduce its coal dependence in the electricity market to around 50% by increasing the use of gas. Here the motivation includes the wish to expand the outlets for indigenous gas.

*Italy*

Until recently, demand for coking coal has been steady, but other non-power markets are not significant. The use of coal for power generation increased to 1991, but has fallen away subsequently. In spite of the now complete absence of nuclear power, coal use has been limited by the continuing high use of fuel oil and natural gas in conventional power plants. Italy has no significant coal production, and imports virtually all its requirements. The main issue for government has been the inability to strike a balance between regional and central plans: deadlock has prevented significant change to the pattern of power generation, and has inhibited coal use. The above projections imply broad stability in power station coal use: but this may prove to be an overestimate if significant increases in gas use take place.

*Portugal*

Apart from cement industry requirements, most coal demand is for power stations, built since 1973, all supplied by steam coal imports. Notwithstanding a more than doubling of electricity demand since 1973, and the absence of nuclear power, the increase in the use of coal has been modest due to higher fuel oil consumption. The projections imply a further modest increase in coal use, with reductions in fuel oil use being taken by gas.

*The Netherlands*

The coal market has more than doubled since 1973. Non-power markets have been stable, but supplies to power stations have shown a large increase, reflected in additional imports. Coal production has ceased. Despite a very limited nuclear power element, the increase in coal consumption has been modest because of the continued high level of use of natural gas in conventional power plants. Although the above projections include a small increase in coal at the expense of gas, policy in this area is uncertain.

*Ireland*

All Ireland's hard coal use is in power generation, and all the coal is imported. Despite a doubling of electricity demand since 1973, use of coal has been limited by use of peat and natural gas. No significant change is expected.

*Finland*

Most Finnish hard coal use is in power generation, and all the coal is imported. Despite a doubling of electricity demand since 1973, use of coal has been limited by nuclear expansion. Future projections imply that coal shares with gas the continuing rapid increase in electricity demand.

*Sweden*

The main uses for hard coal are in iron and steel, power generation and industry, and all the coal is imported. Despite a doubling of electricity demand since 1973, increased use of coal has been limited by nuclear expansion and the high level of hydropower, which together provide well over 90% of electricity output.

*Others*

In Austria, Norway and Switzerland, coal consumption is very limited, due in large measure to the use of hydro-electricity.

## 1.5 The issue of coal-fired generating plant

Over the next decade or so, the capability of western Europe to use coal will in large measure reflect existing coal-fired generating capacity (although inter-fuel competition will continue to affect the loading of this plant). For the longer term, however, the age profile of the current existing plant will become increasingly important. Table 1.12 shows the current position for plants using predominantly hard coal. If the average expected life of coal-fired plant is 30 years, then one would expect to see approaching half of existing plant to have been commissioned since 1980 if the total were to be sustained; yet less than a third of capacity is of this age. The amount and proportion of elderly plant are particularly striking in the UK, but are also marked in France, Belgium and Italy. By contrast, much coal plant in Germany is relatively modern. Although in practice the life of plant can be extended well beyond 30 years (with refurbishment expenditure), in western Europe as a whole the amount of new plant likely to be built in the 1990s (summarized in Table 1.13) will in due course be quite insufficient to offset the obsolescence of plant commissioned before 1970.

**Table 1.12 Existing generating plant using hard coal (capacity in GW)**

| | Date Commissioned | | | | |
|---|---|---|---|---|---|
| | Before 1970 | 1970-80 | 1980-90 | After 1990 | Total |
| Austria | — | — | 1.1 | — | 1.1 |
| Belgium | 1.8 | 1.0 | 0.1 | — | 3.2 |
| Denmark | 2.5 | 2.5 | 1.9 | 1.1 | 8.0 |
| Finland | 0.4 | 1.7 | 0.7 | 0.6 | 3.4 |
| France | 5.1 | 3.2 | 1.9 | — | 10.2 |
| Germany | 9.6 | 6.5 | 13.7 | 2.0 | 31.8 |
| Greece | — | — | — | — | — |
| Ireland | — | — | 0.9 | — | 0.9 |
| Italy | 3.9 | 2.9 | 0.2 | 1.4 | 8.4 |
| Netherlands | 1.0 | 1.8 | 0.8 | — | 3.6 |
| Portugal | — | 1.3 | 0.3 | — | 1.6 |
| Spain | 1.6 | 1.4 | 4.6 | — | 7.6 |
| Sweden | — | 0.1 | 0.6 | — | 0.7 |
| UK | 19.2 | 15.0 | 2.6 | — | 36.8 |
| Total | 45.1 | 37.4 | 29.4 | 5.1 | 117.3 |

Source: *World Coal-fired Power Stations: Europe and Russia*, Paris: IEA Coal Research, 1994.

Given that some of the 'planned' capacity will be late coming into use, or may not proceed at all, the likely total of coal-fired plant commissioned in the 1990s (including new plant already in operation) is unlikely to exceed 10 GW – an inadequate quantity if a progressive fall in capacity at some point after 2000 is to be avoided. Thus, even if the total market for coal over the next decade shows only modest decline, this will undoubtedly accelerate later unless there is a marked increase in investment in new plant.

### 1.6 Emergence of environmental issues

Before summarizing the argument so far and identifying the issues for the future, it is necessary to note the increasing importance of environmental considerations in the market for coal, and the extent of variations in approach between the individual countries within western Europe. There are two main

**Table 1.13 New generating plant using hard coal (capacity in GW)**

| | Under construction[a] | 'Planned'[b] |
|---|---|---|
| Finland | 0.1 | — |
| Germany | 0.4 | 3.8 |
| Portugal | 0.3 | 0.6 |
| Spain | 0.9 | — |
| Denmark | — | 0.4 |
| Netherlands | — | 0.6 |
| Total | 1.7 | 5.4 |

[a]About 1 GW of plant listed as U/C for completion in 1994 now included in 'existing'.
[b]Firm announced plans for commissioning before 2003.
Source: IEA Coal Research.

issues: conventional pollutants ($SO_2$, $NO_x$ and PM) associated with 'acid rain'; and the wider issue of climate change ($CO_2$).

On the related issues of air quality and 'acid rain', the EC introduced a Directive in June 1984 on Air Pollution for Industrial Plant (the 'Framework Directive'), which required the application of BATNEEC (best available technique not entailing excessive cost). This was followed in 1988 by the Large Combustion Plant Directive (LCPD), which calls for a three-stage reduction from 1980 levels of $SO_2$ emissions from existing plant over 50 MW, with overall EU-12 targets of 25%, 43% and 60% reductions by 1993, 1995 and 2003 respectively, and for a two-stage reduction of $NO_x$ with a 20% reduction by 1993 and 36% by 1998. However, different targets were set for individual countries in respect of both $SO_2$ and $NO_x$, to take account of differences in economic, energy and environmental circumstances. The $SO_2$ targets which are the most important for coal are set out in Table 1.14.

The recent agreement on the Second Sulphur Protocol to the UNECE Long-Range Convention on Transboundary Air Pollution added long-term reduction goals across the European UNECE region.[10] This will almost certainly have

[10] The UN Economic Commission for Europe encompasses the countries of western, central and eastern Europe (as well as, formally, the US, which was not involved in the Long-Range Transboundary Air Pollution (LRTAP) Convention). The First Sulphur Protocol to that Convention (in 1985) established the '30% Club' of countries committed to reducing sulphur emissions by 30% by the mid-1990s. The UK was the most prominent non-signatory, but has joined in the Second Sulphur Protocol.

**Table 1.14 Sulphur reduction targets in the Large Combustion Plant Directive (% reduction from 1980 levels for existing large plant).**

| | By 1993 | By 1998 | By 2003 |
|---|---|---|---|
| UK | 20 | 40 | 60 |
| Germany | 40 | 60 | 70 |
| Spain | — | 24 | 37 |
| France | 40 | 60 | 70 |
| Belgium | 40 | 60 | 70 |
| Denmark | 34 | 56 | 67 |
| Italy | 27 | 39 | 63 |
| Portugal | +102 | +135 | +79 |
| Netherlands | 40 | 60 | 70 |
| Luxembourg | 40 | 50 | 60 |
| Ireland | +25 | +25 | +25 |
| Greece | + 6 | + 6 | + 6 |

Source: Author.

the effect of tightening the LCPD targets in the longer term. Although the new limits agreed to $SO_2$ emissions relate to all sources, in practice large existing power stations will have to bear the greater part of further reductions, and this agreement may in fact override earlier proposals to revise the targets under the LCPD. The comparisons set out in Table 1.15 are therefore unlikely to be misleading.

While BATNEEC is a means of providing an extended period of transition for existing plant, it is clear that the ultimate aim is to move towards a situation where 90% desulphurization would be required of all large power plants (500 MW or more). These requirements put coal (and high-sulphur fuel oil) at a significant disadvantage against natural gas and nuclear power, as can be seen from the relative emissions in Table 1.16. Coal's disadvantage is particularly marked using 'conventional' plant without flue gas desulphurization (FGD). But FGD is expensive, and 'clean coal' is not yet proven (see Chapter 4).

The LCPD also provides for more stringent requirements for new plant, effectively requiring 90% desulphurization for plant of 500 MW, ranging down to 40% for plant of 100 MW (from which Spain is exempt to a degree up to 1999). Again, this provision is subject to review in 1995. For new

**Table 1.15 Sulphur reduction targets in the UNECE Second Sulphur Protocol ($SO_2$ emission % reductions from 1980 levels, existing large plant)**

| | LCPD 2003 | 2005 | '2nd Protocol' 2010 |
|---|---|---|---|
| UK | 60 | 70 | 80 |
| Germany | 70 | 87 | |
| Spain | 37 | (no significant change) | |
| France | 70 | 76 | |
| Belgium | 70 | 72 | |
| Denmark | 67 | (80 by 2000) | |
| Italy | 63 | 73 | |
| Netherlands | 70 | (77 by 2000) | |

Source: Author.

plant, most EU-12 countries have adopted the LCPD standard for $SO_2$ emissions of 400 mg/Nm$^3$ for large plant, although standards in the Netherlands and Belgium are more stringent, while Spain and Portugal have no effective controls. For existing plant, Germany and the Netherlands have already adopted standards requiring FGD on all continuing power stations. By contrast, there are virtually no $SO_2$ controls at stations in France, Belgium, Spain, Portugal or Ireland, and only a few stations in Denmark, Italy and the UK have controls.

There is also considerable variation between countries in respect of measures to control emissions of $CO_2$. The European Council declared in October 1989 the objective of stabilizing EU-12 emissions of $CO_2$ in 2000 at their 1990 levels, a target repeatedly reaffirmed by subsequent Council meetings. The target is one for the EU-12 as a whole, and envisages significant variations between countries; it was formulated partly on the basis of national targets adopted (with varying degrees of commitment) by many member states during 1990, as summarized in Table 1.17. The Commission, asked to propose measures to achieve the common goal, put forward proposals for a range of measures including a carbon/energy tax equivalent to $10 per barrel of oil by 2000; but no agreement has yet been reached on these.

**Table 1.16 Relative emissions of $SO_2$ and $NO_x$ from different generating technologies**

| | $SO_2$ | $NO_x$ |
|---|---|---|
| Coal (1.5% sulphur) | 1.0 | 1.0 |
| Coal (0.9% sulphur) | 0.6 | 1.0 |
| Coal with FGD + low $NO_x$ burners | 0.1 | 0.6 |
| High S. Fuel Oil (without FGD) | 1.2 | 0.75 |
| Orimulsion (without FGD) | 1.3 | 0.65 |
| Gas (CCGT) | 0 | 0.25 |
| 'Clean coal' | 0.05/0.1 | 0.3/0.4 |
| Nuclear | 0 | 0 |

Source: Author.

In many ways, the $CO_2$ issue is more intractable than $SO_2$ as a problem for coal, since $CO_2$ is an inevitable result of coal combustion, and coal has a higher carbon content than other fuels (see Table 1.18). Coal's disadvantage can be mitigated only by higher thermal efficiencies (such as 'clean coal' technologies which incorporate the combined cycle). Thus the emergence of $CO_2$ as a policy issue in the late 1980s has weakened the strategic position of coal and raised the prospect of higher costs/taxes relative to natural gas or nuclear power.

## 1.7 Issues for the future

There are very considerable variations in the structure of the coal market between the individual countries in western Europe, both as a result of different economic and political influences and also because of different patterns of energy production and use 'inherited' from the past. Because of the capital intensity of energy use and the often difficult social issues associated with coal production, these patterns are not usually susceptible to very rapid change. Nevertheless, certain themes have emerged.

First, there is the issue of high-cost indigenous coal production. It is notable that EU-12 coal production has declined substantially since 1973, notwithstanding the large measure of subsidy and protection the EU-12 coal industries

**Table 1.17 $CO_2$ emission reduction targets in the EU-12**

| Country | Target |
|---|---|
| Belgium | -5% by 2000 (cf. 1990) |
| Denmark | -20% by 2005 (c.f. 1988) |
| France | 2 tons carbon/per capita by 2000 |
| Germany | -25% by 2005 (cf. 1987) |
| Greece | None yet |
| Ireland | No more than +20% in 2000 (cf. 1990) |
| Italy | 'Stability', but within framework of EU-12 target |
| Netherlands | 3-5% by 2000 (cf. 1989/90) |
| Portugal | None |
| Spain | No more than +25% in 2000 (cf. 1990) |
| UK | No increase by 2000 (cf. 1990) |

Source: Author.

have enjoyed and the fact that, for much of the period, oil prices were very high. Rather, the markets for EU-12 hard coal have been reduced by the build-up of low-cost imported coal from outside the Community, by the expansion of nuclear power and (more recently in the UK) by the entry of gas into power generation via the use of CCGTs. Following the fall in oil prices in 1986, and the continuing oversupply of energy markets, 'security of supply' arguments in favour of indigenous coal production have looked increasingly weak; and, in any event, the majority of member states in the EU-12 (and other OECD countries in western Europe) have no reason to support high-cost European coal: they want unfettered access to cheap internationally traded coal. It is therefore necessary to examine the policies that have been pursued by the Commission and the relevant member states on the issue of subsidy and protection, and to consider how these policies are likely to develop in the circumstances of the 1990s. This will concern notably the UK, German and (to a lesser extent) Spanish governments, and the Commission.

Secondly, there is the question of the coal market in the coal-producing countries, irrespective of its source, and how this might be affected by wider developments in western Europe. As we have noted, in all those countries where there has been a net increase in the coal market since 1973, this has been based exclusively on imports (except in the case of Spain, but even here,

**Table 1.18 Relative emissions of $CO_2$ from different generating technologies**

| Fuel | Index[a] |
|---|---|
| Coal (conventional) | 1.0 |
| Coal (with FGD) | 1.05 |
| Heavy Fuel Oil | 0.85 |
| Orimulsion | 0.85 |
| Gas (CCGT) | 0.5 |
| 'Clean Coal'[b] | 0.7/0.9 |
| Nuclear | nil |

[a] Coal = 1 per unit electricity generated.
[b] The use of combined heat and power schemes reduces emissions further, but the relative disadvantage of coal remains.
Source: Author.

most of the increase in coal use was attributable to imports). Yet, in spite of the very large cost advantage imported coal had over fuel oil for most of the period, and the strong policy pressures to use coal to reduce dependence on OPEC oil, the impact on the coal market was very uneven, and modest in extent: the increase in coal use in those countries where the coal market did expand was equivalent to little more than 2% of total western European energy demand. For much of the 1970s and 1980s, the market for coal was constrained by the expansion of nuclear power in a number of countries. On the other hand, the view that natural gas was a 'noble fuel', too valuable to be used in power generation, and the rapid growth of electricity demand in some countries, provided opportunities for coal.

However, these factors have begun to work in the opposite direction. Moreover, it has been noted that, in western Europe as a whole, the introduction of new coal-fired generating capacity in the 1990s is likely to fall short of the quantities required to replace old plant beyond the turn of the century. Future coal use will be influenced greatly by environmental concerns which, other things being equal, tend to work to the disadvantage of coal, particularly if countervailing concerns for energy security are weak in the future energy market. Although the impact of 'environmental politics' on coal will vary greatly between countries, few countries are likely strongly to defend coal's long-term position unless there is a major breakthrough in the economics of 'clean coal' technology.

In the resolution of these issues – the protection of high-cost production, the evolution of the coal market and the impact of environmental regulation – politics will play as important a part as, if not a more important part than, economics; and politics will also decide the extent to which these issues will be resolved by collective action, or by action by individual governments.

# Chapter 2

# The Problem of High-Cost Production

This chapter examines the economics and politics of protecting higher-cost European coal production against cheaper imports and other alternatives.

## 2.1 The extent of uncompetitiveness

The issue of subsidy is by no means clear-cut, since effective 'protection' for high-cost production may be provided by means other than direct grants raised by taxation. Distortions of the market, in particular imposed contracts or other sales arrangements designed to provide revenues to coal producers higher than would be obtained in competitive markets, are also relevant. On the other hand, some grants given to coal industries have been to facilitate rationalization and rundown, and therefore fall in a different category from subsidies designed to support operations on a continuing basis.

The IEA has attempted to address this problem with the use of the producer subsidy equivalent (PSE), employing a methodology previously used to make comparisons on agricultural products. A PSE represents the payment that would just keep all domestic production competitive with imports, at existing levels of coal output and current import prices, in a particular year. As the IEA itself emphasizes, the methodology is static, and cannot indicate the savings that could be realized immediately if protected production were closed down (since all European production could not be closed down without increasing the import price). The PSE includes the following elements:

- net budgetary payments to producers designed to sustain current production, but excluding direct payments designed to speed contraction or otherwise unrelated to current production;
- the difference between actual prices of indigenous production (including prices under contracts supported by governments) and the level of international prices, with some adjustment for transport costs.

**Table 2.1 Trend in producer subsidy equivalents for hard coal in Europe (1990 $US)[a]**

| | 1986 | | 1990 | | 1991 | | 1992[b] | |
|---|---|---|---|---|---|---|---|---|
| | $ mill | $/tonne | $m | $/tonne | $m | $/tonne | $m | $/tonne |
| Germany | 4996 | 62 | 6945 | 99 | 6939 | 105 | 6828 | 105 |
| Spain | 447 | 20 | 634 | 33 | 1002[c] | 55 | 656[c] | 36 |
| UK | 3311 | 31 | 2418 | 26 | 1842[c] | 19 | 1719[c] | 18 |

[a] There are no figures for France as it was not then an IEA country.
[b] Provisional.
[c] IEA estimate.
Source: *1992 Review of Energy Policies in IEA Countries*, (Paris IEA/OECD, 1993).

Table 2.1 shows the evolution of the PSE in Germany, Spain and the UK in constant 1990 US dollars. By 1992 the overall level of protection in Germany, already high in 1986, had further increased and the level of support in Spain was higher than in 1986, whereas there had been some reduction in the UK. Most of the PSE for the UK was the computed difference between international prices and the prices in the contracts between British Coal and the electricity generating companies National Power and PowerGen, rather than direct subsidy. This evolution has occurred against a background of falling 'real' prices of coal imported into the EU-12 from third countries (Table 2.2).

International comparisons of costs and revenues are always difficult, because of differences in accounting conventions, treatment of social security payments and exchange rates (which often diverge greatly from long-term equilibrium levels). In the case of coal, there are in addition significant variations in coal quality between and within countries, and in transport costs (which affect the economic value of production in competition with imports). Nevertheless, using Commission figures some comparisons can be made. Table 2.3 compares the main EU-12 coal producers, using 1992 exchange rates.

In 1992, coal imported into the EU-12 from third countries had an average CIF price of around 1.4 ECU/GJ. Because of internal transport costs and other factors, the value of EU-12 production in competitive terms might have been somewhat higher than that – say 1.6 ECU/GJ. If 1.6 ECU/GJ = 100, then the cost relativities were as follows:

**Table 2.2 Trend of imported coal prices**

| | Coking coal | | Steam coal (power stations) | |
|---|---|---|---|---|
| | US/tce (actual) | US/tce 1992 $US | US/tce (actual) | US/tce 1992 $US |
| 1986 | 55.1 | 63.5 | 48.2 | 55.5 |
| 1990 | 56.7 | 57.7 | 54.2 | 54.6 |
| 1991 | 56.1 | 56.3 | 52.0 | 52.2 |
| 1992 | 54.5 | 54.5 | 52.0 | 52.0 |
| (1993) | (52.9) | | (47.2) | |

Source: IEA Coal Information 1993 ed, tables 2.3, 2.4.

| | |
|---|---|
| Spanish deep mined | 437 |
| German deep mined | 305 |
| French deep mined | 213 |
| Spanish opencast | 162 |
| UK deep mined | 161 |
| UK opencast | 114 |

These figures show the huge size of the gap between the cost of EU-12 deep-mined coal and international coal prices in 1992. UK costs have fallen substantially since 1992 (unit costs fell by 13% in 1993/4), but for Germany and Spain the size of the gap remains the basis of the politics of coal in these countries. The figures also show the extent to which (even before recent further dramatic improvements) UK deep-mined coal was by far the lowest-cost and most productive in the EU-12.

It is instructive to note the relative manpower and productivity changes in 1992 and 1993 (Table 2.4). In Germany, there was a small (1%) increase in output *per man-shift* in 1993 over 1992, but this was offset by a reduction in working days in order to reduce output in the face of overproduction; in the UK, by contrast, the fall in deep-mined output was associated with a further large manpower reduction and productivity increase. Between the first quarter of 1992 and the fourth quarter of 1993, underground manpower fell by an astonishing 66% in the UK, compared with 10-20% in other European coal-producing countries.

**Table 2.3 Volume, cost and productivity of coal production in Europe, 1992**

| | | Output (million tonnes) | Average costs (ECU/GJ) | Output per man/year (underground) (tonnes) |
|---|---|---|---|---|
| Germany | deep mined | 71.8 | 4.88 | 916 |
| France | deep mined | 8.3 | 3.42 | 1107 |
| UK | deep mined | 61.4 | 2.58 | 1840 |
| | opencast | 15.0 | 1.83 | — |
| Spain | deep mined | 13.8 | 6.99 | 483 |
| | opencast | 4.9 | 2.60 | — |

Source: European Commission.

## 2.2 The subsidy regime to 1992

Coal policy in the EU-12 falls within the framework of the European Coal and Steel Community (ECSC) Treaty (the Treaty of Paris) rather than the Treaty of Rome. Article 4 of the ECSC Treaty states that 'subsidies or aids granted by States, or special charges imposed by States, in any form whatsoever' are 'recognised as incompatible with the common market for coal and steel, and shall be accordingly abolished and prohibited within the Community'. This is starkly at variance with current permitted practice. However, circumstances are now very different from those at the time the ECSC Treaty was agreed in 1951, when coal constituted the major part of Community primary energy supply (other than oil for transport fuels and hydroelectricity), and was in short supply. In these circumstances, those wishing to create a common market in coal and steel were concerned to avoid subsidies which would give one country's industry an unfair advantage over another's and impede the development of intra-Community coal trade.

By the 1960s, this rationale was undermined by the influx of cheap Middle East oil. The then Community coal industries (in Germany, France, Belgium and The Netherlands) were forced to contract, and needed operating subsidies to cover the difference between coal costs and oil prices. This was recognized by the Commission, which invoked the exceptional provisions laid down in

**Table 2.4 Productivity and labour changes in European coal production, 1992-1993**

| | Output per man year (underground) tonnes per man | | | Labour force ('000) | | |
|---|---|---|---|---|---|---|
| | 1992 | 1993 | % change | 1st quarter 1992 | 4th quarter 1993 | % reduction |
| UK | 1785 | 2335 | +31% | 37.8 | 13.0 | (66) |
| France | 1103 | 1060 | -4% | 8.7 | 7.9 | (10) |
| Germany | 916 | 889 | -3% | 72.4 | 62.0 | (14) |
| Spain | 454 | 544 | +20% | 29.7 | 24.0 | (19) |
| Other | | | | 2.3 | 0.5 | |
| EU-12 Total | | | | 150.9 | 107.4 | (29) |

Source: Eurostat *Rapid Report*, 1994/25, 1994/26.

Article 95 of the ECSC Treaty to give effect to the State Aids Decisions of 1965 and 1971, which allowed subsidies to enable the coal industries to adjust to lower levels of production and to mitigate social problems in coal-mining areas. Following the OPEC price revolution of 1973-4, the 1976 Decision (which was valid for 10 years) no longer confined its aim to alleviating social problems arising from contraction, but also sought to stabilize Community coal production under satisfactory economic conditions. However, by the time the replacement Decision (Decision No. 2064/86/ECSC) came into force in 1986, the previously high oil prices had collapsed, the European gas market had been greatly developed, the nuclear power programme was having a major impact and imports of low-price coal from outside the Community were increasingly available. The new Decision (which was due to operate until 1993) abandoned the objective of stabilizing Community coal production, and laid down that state aids to the coal industry had to satisfy one or more of the following tests, that they would help to:

- improve the competitiveness of the Community coal industry, thereby contributing to security of supply;
- create new capacities, provided that they were economically viable;

- solve the social and regional problems associated with coal industry restructuring.

The Commission produced a 'mid-term' report on the application of the 1986 Decision (SEC(91)1983 FINAL) in which it expressed dissatisfaction with the way the Decision had worked. It drew a distinction between Germany and Spain on the one hand, where state aids had been used to sustain high-cost industries, and the UK and France on the other hand, where they had been used primarily to assist restructuring (as intended under the terms of the Decision). The Commission considered that it would be particularly unfair for relatively low cost pits to continue to close in the UK, while heavily subsidized high-cost production in Germany and Spain continued to operate. The Commission reaffirmed its view that 'production as such, irrespective of economic viability, cannot be considered an end in itself,' and it was clearly impatient with those state aids that were 'sustained by arguments based on security of supply, 'since security of supply is not guaranteed by producing coal in the Community'.

The report by the Commission on the application of Community rules for state aid to the coal industry (COM(93)589FINAL) reviewed in detail the application of state aids for coal mining and restructuring in 1992. Table 2.5 presents these data, excluding Belgium (which has now ceased production) and Portugal (where production is insignificant). These figures serve to emphasize the great divergence in approach between member states. In the UK there was no direct subsidy by 1992, though direct support through long-term contracts was important, as noted below. In the other producing countries, the 1992 arrangements in support of current production can be summarized as follows.

*Germany*

Germany has the most complex system of state aid in the Community.

- The deficit grants under Article 3 were to cover revenue shortfall in certain mines sending low volatile coal to power stations, though this is to be phased out.
- The sales aids under Article 4 were entirely for the sale of coking coal and coke to the steel industry, representing the difference between prices aligned to international levels and costs of production, up to a maximum production

**Table 2.5 State aids for coal production and adjustment in Europe (1992)**

| | Germany | | France | | Spain | | UK | |
|---|---|---|---|---|---|---|---|---|
| | Million ECU | ECU/t | Million ECU | ECU/t | Million ECU | ECU/t | Million ECU | ECU/t |
| *Direct aid* | | | | | | | | |
| Article 3 | | | | | | | | |
| Definite grants | 178 | 2.4 | 187 | 19.7 | 380 | 20.4 | | |
| Article 4 | | | | | | | | |
| Sales aid | 1788 | 24.8 | — | — | — | — | | |
| Article 5 | | | | | | | | |
| Investment aid | — | — | — | — | — | — | | |
| Article 6 | | | | | | | | |
| Aid for underground miners | 65 | 0.9 | — | — | — | — | | |
| Other | — | — | — | — | 66 | 3.5 | | |
| Sub total | 2031 | 28.1 | 187 | 19.7 | 446 | 23.9 | nil | nil |
| Indirect aid | 2467 | 34.2 | — | — | 17 | 0.9 | — | — |
| Total | 4498 | 62.3 | 187 | 19.7 | 463 | 24.8 | nil | nil |
| *Financial aids not related to current production* | | | | | | | | |
| Article 7 | | | | | | | | |
| Social benefits | 4436 | | 2364 | | 858 | | nil | |
| Article 8 | | | | | | | | |
| Inherited liabilities from restructuring | 247 | | 775 | | 20 | | 13 | |

Note: Articles referred to are those of Decision No 2064/86/ECSC.
Source: European Commission (93)589 FINAL, Brussels, November 1993.

level of 20.7 million tonnes in 1992 (although this is planned to decline to 15 million tonnes).

- The aids under Article 6 were to pay for miners' shift bonuses on the grounds that this improves the competitiveness of the industry.
- The indirect aid represented payments to the coal industry under the Third German Electricity-from-Coal Law, representing the recycling of the levy (the *Kohlepfennig*) raised from electricity prices to cover the difference between the cost of using German hard coal and that of using imported coal and oil.

*France*
Aids were entirely related to covering losses, after applying prices at international levels.

*Spain*
Aids were mainly related to covering losses, but in addition there was aid to cover depreciation and other costs arising from restructuring.

The issue of subsidies cannot be dissociated from policy on coal *prices*, and hence from the basis of coal producer revenues. In the UK, prices in the dominant electricity market until 1993 were determined by long-term contracts on terms agreed with the government, with prices trending towards international levels but well above them in the interim. This difference between world prices and UK prices was not formally a 'subsidy', as no public expenditure was incurred and there was no separate surcharge on electricity prices. However, the effect was to protect the UK industry over a transitional period. In France, prices of electricity coal supplied under contract are related to international prices plus a premium of about 10%. In Germany, coking coal prices are related to international prices, whereas electricity coal prices sold under the *Jahrhundertvertrag* are set to cover the costs of production. In 1992, 87% of German power station coal demand was met by German coal, most of which was sold at high contract prices. Overall, 62% of power station coal was sold at contract prices which were nearly three times the price of imported coal. In Spain, deep-mine coal producers have long-term contracts under which coal is sold at a 'reference price', which is substantially higher than world prices, plus a surcharge to cover higher production costs above the reference price, recovered from a levy on electricity tariffs. The resulting variations in average revenues are summarized in Table 2.6. These variations in what constitutes producer revenue clearly affect the nature and extent of subsidy in particular countries and markets. This is reflected in the difference between the subsidies for current production, as notified by the Commission, and the PSE as calculated by the IEA (see section 2.1 above), also summarized in Table 2.6.

It needs to be emphasized that none of the subsidies or other forms of state aid being considered here come from Community funds. (The coal regime is

**Table 2.6 Variation in revenue PSE and state aids in Europe**

| | Average Revenues (ECU/tce) | 'Subsidies' 1992 (ECU/t) | |
|---|---|---|---|
| | | PSE basis | State aids basis |
| Germany | 127 | 81 | 62 |
| Spain | 107 | 28 | 24 |
| UK | 71 | 14 | — |
| France | 58 | n/a | 20 |

Source: European Commission and Author.

totally different from that applied to agriculture.) All subvention is provided by the member states concerned. The role of the Council of Ministers and Commission is to lay down the rules under which member states may protect their own coal industries.

We now examine the reasoning whereby the Commission justified authorization of the wide variety of practices within the coal-producing states in 1992, as set out in COM(93)589 FINAL, against the background of constant or increasing aids per tonne for current production in the EU-12. The Commission justified authorization in the following terms.

*Germany*

As coking coal aid has been given under a new framework which sets a ceiling on aid for the period 1992-4, it is justified as providing 'a temporary solution to the social and regional problems by spreading out the closure of loss-making pits in time, and also the restructuring of the industry'. On electricity coal aid (via the *Kohlepfennig* levy), the Commission said that in giving authorization it took account of 'the social and regional context typifying the coal sector in Germany, and the programmed moves towards restructuring which should ultimately lead to reduction in the said intervention'. Although the *Jahrhundertvertrag* had been authorized by the Commission to continue in operation to the end of 1995, the Commission made it clear that the current form of state aid would cease to have legal effect beyond the end of 1993, when Decision No. 2064/86/ECSC expired.

*Spain*

Aid to cover operating losses was justified on the grounds that it 'facilitates the process of restructuring in particular by staggering the closure of certain pits with no prospects of economic viability, in the context of regional industrial development policy. The aid thus helps to resolve the social and regional problems related to developments in the coal industry.' The Commission also noted that the Spanish government had also set up a fund to defray part of the specific depreciation and other exceptional costs arising from future restructuring.

*France*

Aid to cover operating losses was justified on the grounds that it 'forms part of a restructuring and rationalization plan that is intended to improve the competitiveness of the industry'.

Overall, the Commission concluded that, although there were problems of aid transparency, 'the results obtained in certain Member States and the decisions recently taken, particularly in Spain and Germany, seem to constitute a major, apparently decisive, step forward towards a healthy restructuring of the Community coal industry'. It is against this background that we next consider the introduction of the new state aids regime.

## 2.3 The new decision on state aids: the problems of agreement

The new direction on state aids to the coal industry (Decision No. 3632/93/ECSC) will operate from the beginning of 1994 until 2002 (when the ECSC Treaty itself expires). The negotiations leading up to the issue of the new Decision were protracted and difficult.

In an effort to find a formula likely to be acceptable to all parties (but particularly the Germans), the Commission floated the idea of a 'reference price'. This was first raised in the 'mid-term report' (SEC(91)1983)FINAL) in 1991: 'In a future aid system a reference price, expressed in ECU, could constitute an acceptable ceiling and would integrate, at the same time, the regulatory [ie. restraining] effect on the world market of Community production, and, implicitly, the security premium linked to the provision of supplies which are as stable as possible in geopolitical terms.' In effect, the 'reference price' would not be a price at all, but a reference *cost limit* for subsidy purposes, comprising three elements:

- view on the long-term 'sustainable' price of traded coal delivered into Europe;
- view on the additional economic value of Community coal as a restraint on the price which would otherwise be paid for both existing and potential coal imports;
- 'security of supply' premium.

Whatever the merits of this concept, it was overwhelmed by the problems of acceptable quantification. Although there was no formal proposal by the Commission, discussions initially took place on the basis of a 'reference price' of about 90 ECU/tce: about double the import price. While it was possible to argue that the long-run import price might be somewhat higher than current levels, and that the economic value of 'world price restraint' might be up to 10 ECU/tce, this still left a huge margin of perhaps 30 ECU/tce attributable to the value of 'security of supply'. However, although a figure as high as 90 ECU/tce could scarcely be justified as a measure of the economic value of Community coal, this number was far too low to be acceptable to the Germans, who argued for a figure in excess of 120 ECU/tce, and even then were unwilling to accept that this could be used as an absolute upper limit on subsidizable costs. They also opposed a later Commission idea to limit subsidies to output with costs below a reference price related to average Community production costs. The fact was that there was no single formula which could both provide a justification for the continuing levels of German subsidies and, at the same time, bear a sensible relationship to the long-term economic value of Community coal production.

By the same token the 'reference price' concept provided no escape from what may be termed the Anglo-German anomaly. As the Commission put it in the 'midterm report': 'It would be particularly unfair for pits to continue to close in those member states which have nearly completed the restructuring exercise, while substantially less productive pits in other coal-producing member states were left running thanks to a national system of subsidies and other protective measures.' Thus, while the Germans were refusing to accept that subsidies should be limited to collieries producing at less than 90 ECU/tce, the UK industry was faced with having to close collieries capable of producing at half that level of cost, as subsidies were always regarded as permissive rather than obligatory, and the German government wished to subsidize, but the UK government did not.

While the 'reference price' concept sought to apply an upper limit on the costs attracting subsidy, there remained the problem of the *quantities* to which subsidies should be allowed to apply. These matters had been regarded as outside the powers of the Commission, since Article 71 of the ECSC Treaty states that 'the powers of the Governments of member states in matters of commercial policy shall not be affected by this Treaty, save as otherwise provided therein'; and Article 73 states that 'the administration of import and export licences for trade with third countries shall be a matter for the government in whose territory the place of destination for imports is situated'. These provisions had been used by the Germans to justify quantitative restrictions on coal imports. The Commission attempted to address this issue in a Working Paper in 1990 following agreement between Energy Commissioner Cardoso and Competition Commissioner Brittain.[11] The Commission intended to ensure that the proportion of primary fuels supplied for electricity generation in each member state which was protected by special measures such as long-term supply agreements or aids did not exceed the levels already attained. The final version of this paper forwarded to the Council of Ministers and the ECSC Consultative Committee did not quantify the 'maximum level of national protection', but earlier versions put the figure at 20%, falling to 15% by the end of the century, and these figures continued to be used by the Commission in policy discussions. The timing of the original proposition in mid-1990 was clearly influenced by the need to rationalize the UK nuclear levy and the associated 'non-fossil fuel obligation' as part of the electricity privatization settlement (to which the Commission had given agreement with considerable reluctance). Twenty per cent was just about the nuclear primary energy input in electricity generation in the UK; and in Germany the same figure could be applied to indigenous hard coal. But the

---

[11] *European Commission, Security of Supply: The Internal Energy Market and Energy Policy* (SEC(90)1248FINAL), Brussels, September 1990.

'20% rule', used in the UK to support nuclear energy, but in support of hard coal in Germany, only served further to entrench the Anglo-German anomaly so far as coal was concerned. This was effectively recognized by Commissioner Cardoso when he gave evidence to the House of Commons Trade and Industry Select Committee in December 1992 at the time of the UK 'coal crisis'.[12]

The heart of any opposition to significant weakening of the state aids system came from Germany. As had been the case in the past, the Germans proceeded by consent and agreement among all parties concerned. In order to defend the position of the German coal industry, in November 1991 an agreement entitled Kohlekonzept 2005 (or Coal Plan 2005) was reached between the federal government of Germany, the state governments of North-Rhine/Westfalia and Saarland, the German coal-mining companies (principally Ruhrkhole AG) the Mining and Energy Industrial Trade Union, the Association of Electric Power Companies and the Association of Industrial Power Suppliers. The preliminaries to this agreement had been:

- a 'decision of general principle' reached in 1989 between the Federal Chancellor and the prime ministers of North-Rhine/Westfalia and Saarland that 41 mtce per annum would continue to be applied to power stations until the end of 1995, under the terms of the *Jahrhundertvertrag* (although the mining industry would give up previously projected increases in protected tonnage);
- a report by the 'Coal Commission' (under Professor Mikat) designed to set out the 'optimum technological and economic utilization of capacity'.

The main terms of the Kohlekonzept 2005 agreement were as follows:

- The *Jahrhundertvertrag* contract remained unchanged, with 41 mtce per annum of protected output being sold to power stations to the end of 1995. 'Subsequent contracts should guarantee the use of German hard coal in power plants.' These tonnages would then gradually decrease to 35 mtce by 1997 and continue at that level until 2005.

---

[12] House of Commons Trade and Industry Select Committee, *British Energy Policy and the Market for Coal*, London: HMSO, 1993, para. 238.

- 'The validity of agreements securing indigenous coking coal supplies for the German steel industry should be due extended until the end of 2005.'
- The total volume of protected coal for the electricity and coking industries would be reduced to 50 million tonnes by 2005, and probably by 2000. (In addition, there would be some more limited tonnages in the domestic and heating markets.)
- The mining companies would focus production on the most efficient sites, and reduce the number of mines from 26 to 17, entailing an additional reduction (over and above existing plans) of production by 10 million tonnes a year by 1995 and a further 4 million tonnes a year thereafter (compared with 66 million tonnes in 1991). There would be a loss of 40,000 jobs over the period (compared with the 123,000 employed at the end of 1991).

It is interesting to analyse the justification put forward for this agreement, namely:

- A 'cost-optimized' German coal mining industry is 'imperative' in order to 'guarantee security of energy supply', particularly as 'coal is the only domestic energy source of any significance in the Federal Republic of Germany'. (This overlooks the distinction between hard coal, which is heavily protected, and brown coal/lignite, which is not.)
- Coal policy is designed to limit dependence on imported sources of energy, particularly by fuelling 26% of power generation with indigenous hard coal. The 35 mtce in the agreement is the minimum, since this would represent less than 20% of power station fuel by 2005.
- German coal makes a contribution to EC security of supply: 'The decisive criterion in the internal energy market is the security of supply against which a future European coal and aid policy is to be measured.'

The emphasis on 'security of supply' in large measure rests on the view that the European Commission has no legal competence in this area. As the Kohlekonzept 2005 document says: 'The ensuring task of the Commission was and remains to verify whether coal aids granted at national level distort competition within the common market or are compatible with the "proper functioning" of the common market. The Commission has no further powers under the Community Treaties: it cannot exercise competence in the field of energy policy. Such competence rests with the national political authorities, the elected parliaments and governments.' The German view is that, on this basis, there is no difficulty with continuing the German state aids system, as

- 'national aid to coal production does not harm the interest of other member states';
- 'no Community resources from the EC Commission or resources from other member states need be used in pursuing the energy objectives of the Federal Republic of Germany'.

In this way, the German coal industry and German government presented a united front against attempts by the Commission to erode the subsidy regime.

One of the chosen vehicles for discussions leading to the new state aid Decision was CEPCEO (the Western European Coal Producers' Association). Here, the 'Anglo-German anomaly' to which we have referred took a different form. The centre of the problem was that the German industry insisted that the very high levels of protection currently available to them should continue, and were in large measure supported in this by the German government. On the other hand, British Coal wanted to create a profitable business without subsidy but buttressed by long-term coal contracts containing an element of price premium, requiring some, but not excessive, contraction in output; while the UK government, in allowing (and indeed encouraging) the 'dash for gas', did not support BC's position at that time. The German coal industry regarded the aim of becoming competitive as a snare and a delusion, and indeed impossible in their case; any contraction had to be modest and phased over a period with full agreement. When it became clear, by 1991-2, that BC faced a rapid halving of its market, the German coal industry was almost incredulous.

This situation in turn was reflected in totally different attitudes to the Commission: for the German coal industry, the Commission was the 'enemy', which wished to undermine the German coal subsidy and protection system, while their own government was in large measure their ally. For British Coal, the Commission was a possible ally in seeking to mitigate the rigours of the UK government's policy towards coal, in order to avoid the contraction of the

Community deep-mine coal industry being concentrated on the lowest-cost producer. In spite of this confusion of purpose, because it was unrealistic for BC (a state-owned industry) to take a line in Brussels which was unacceptable to the UK government, effectively CEPCEO representations were in support of the German position, which was also helpful to the Spanish industry, and not unacceptable to the French.

## 2.4 The new Decision on state aids: terms and conditions

In the preamble to the new state aids Decision (No. 3632/93/ECSC), the Commission set out the justification for continuing with state aids, in spite of their prohibition in the ECSC Treaty:

- 'Most coal production in the Community remains uncompetitive *vis-à-vis* imports from outside the Community,' but the scope for becoming competitive is 'limited by unfavourable geological conditions', so that 'more financial resources are needed than the undertakings themselves can provide.'
- 'Despite the inevitable restructuring and closures, care must be taken to minimize the social and regional impact of these changes, when continuing the Community's policy in this sector, which must take account of the precarious social situation of mining regions.'

The Commission concluded that 'the Community is therefore confronted with a situation for which no provision is made in the [ECSC] Treaty but on which action must nevertheless be taken,' and as the Community 'does not have at its disposal the resources needed to finance this process', aids must continue to be a matter for member states.

It is clear that the Commission's rationale for extending the regime for state aids is effectively designed solely to mitigate the social problems of the contraction of uneconomic capacity, not to argue for its continuance on strategic or 'energy policy' grounds. Indeed, the Decision dismisses the 'security of supply' argument for protection (which had been mentioned in the 1986 Decision) on the grounds that:

- 'The world market in coal is stable with abundant supplies from a wide variety of geographical sources, with the result that even in the long term and with increased demand for coal the risk of persistent interruption of supply, although it cannot be ruled out totally, is nevertheless minimal.'
- Most of the coal imported into the Community comes from IEA countries or states with which the Community or member states have signed trade agreements, and which cannot be considered high-risk suppliers.

It is also clear that the Commission regards state aids as essentially transitional, that they 'must cause no distortion of competition; and that aid should be transparent'.

These principles are reflected in the provisions of the new Decision, which can be summarized as follows:

- 'To make, in the light of coal prices on international markets, further progress towards economic viability with the aim of achieving digression [i.e. progressive reduction] of aids'; 'to solve the social and regional problems created by total or partial reductions in the activity of production units'; 'to help the coal industry adjust to environmental protection standards'.
- Subject to compatibility with the above objectives, and prior authorization by the Commission, operating aid can be given to cover the difference between production costs and 'the selling price freely agreed between the contracting parties in the light of the conditions prevailing on the world market', provided that member states wishing to grant operating aid submit to the Commission in advance 'a modernization, rationalization and restructuring plan' which 'will provide for appropriate measures and sustained efforts to generate a trend towards a reduction in production costs at 1992 prices, during the period 1994 to 2002'.
- State aid can be granted to cover the exceptional costs arising from modernization or restructuring, research and development and 'adjustment to new environmental protection standards'.
- By the end of 1996 at the latest, aids will be authorized only if they are entered in member states' national, regional or local public budgets, and

any existing arrangements linked to agreements between producers and consumers must be modified accordingly.

The Decision bears the hallmarks of compromise between the Commission's rationale and opposition from the Germans. In particular, there is no guidance given on the extent or rate of cost reductions required to satisfy the precondition of aid; the phrase 'sustained efforts to generate a trend towards a reduction in production costs' is of studied vagueness and incapable of precise monitoring. All the earlier ideas of a precise reference price/cost upper limit on output that could be subsidized have been dropped. On the other hand, the requirement that coal prices should bear a reasonable relationship with international price levels, and that the aid should be transparent subsidy in public budgets (and not concealed in the price or provided by levies) will require a renegotiation of the *Jahrhundertvertrag* and the *Kohlepfennig*, which currently provide a large part of the support for the Germany coal industry.

One can also detect in the nature of this compromise some movement from the Commission's earlier position, where there was greater emphasis on removing the evident market distortions represented by heavily subsidized hard coal production in the Single Market. The Commission now appears to have adopted a more relaxed (and realistic) approach on such subsidies, particularly as there is very little intra-Community trade of domestic production (and in the future will be even less), and as the policy of subsidizing German and Spanish coal output in part through higher electricity prices does not confer competitive advantage on the countries concerned, and still remains compatible with the EU-12's environmental objectives. Although Article 3(b) of the Maastricht Treaty, which sets out the subsidiarity principle, is not mirrored in the ECSC Treaty (for which no similar amendment was made) for the energy sector as a whole the subsidiarity principle still appears to leave 'security of supply' judgements with the member states. Moreover, the European Parliament has taken the view that member states retain exclusive competence to organize their national energy markets against this background, and notwithstanding the wide range of interpretation on the way the subsidiarity principle applies in the energy sector, the Commission was not in reality in a position to impose on member states (in this case primarily Germany) a state aids Decision which they would be unwilling to accept. Nevertheless, the Commission has achieved two of its objectives, namely to make subsidies more transparent and to require that they be provided from public budgets rather than as surcharges on prices, which will have to be broadly aligned with international prices.

# Chapter 3

# The Future of Protection in Europe

## 3.1 Introduction: the EU-12 context

We have seen that, notwithstanding the efforts of the Commission, there remains in place a framework for state aid to the coal industry which is not only discretionary and permissive, but which is by no means rigidly drawn. Unlike agriculture, in which all member states have an interest in the issue of protection, only a minority of member states have an interest in domestic coal policy; and even between coal-producer states, there is a wide range of circumstance and policy. It is therefore essentially a policy area ideally suited to the application of the subsidiarity principle. For the non-coal-producer states, the only interest in the policy for Community coal production is to ensure that protective measures do not limit their ability to import freely from third countries. Objections to protectionist measures in favour of EU production have come, not from the other members of the EU-12, but from coal exporters from outside the Union (particularly the US and Australia), which have found their market opportunities in western Europe more limited than they would otherwise have been. In practical terms, therefore, the politics of high-cost EC coal production will continue to be matters for the individual governments of the coal-producer countries, albeit with the Commission acting as 'regulator of last resort' in matters of competition policy.

What policy the coal-producer member states will follow on the protection of high-cost output will depend primarily on:

- their view of the future of the international coal market in terms of security of supply and price;
- their view on the extent to which their coal industries can become competitive without subsidy;
- the politics of social and regional problems in coal-mining areas;
- the politics of trade-offs with environmental concerns.

Environmental issues, which have a large international and Community-wide dimension, are dealt with in Chapter 4 of this study. The other three issues noted above are considered in this chapter with reference to individual coal-producer states.

### 3.2 The future of protection: the United Kingdom

The view of the UK government on the future of the international coal market is not dissimilar from that set out by the Commission in the preamble to the new state aids Decision. In the March 1993 White Paper *The Prospects for Coal*, the government said that 'the diversity of coal reserves and exporters makes it unlikely that political factors or a cartel will in future pose a threat to supplies,' and 'a substantial sustained price increase seems unlikely, given excess steam coal export capacity'.[13] In the first annual *Energy Report* it says: 'The arrival of new supplies on the market will make it likely that there will be a more than adequate supply over the coming years and this is likely to keep prices low'.[14] In other words, the UK Government has taken the view that there is no case for keeping high-cost coal production in the interest of secure and economic coal supply in the future, and that the drive to competitiveness should be concentrated upon reductions in costs and the elimination of uneconomic capacity, rather than awaiting an upturn in international coal prices (or a collapse of sterling against the US dollar).

The interplay of political and economic factors in UK government coal policy to 1993 has been set out in *The UK Coal Crisis: Origins and Resolution*.[15] Since then, legislation for the privatization of British Coal has been passed and its sale put in hand. Deep-mined output has fallen from 62 million tonnes in 1992/3 to about 30 million tonnes by mid 1994. The Preliminary Information Memorandum for the sale of British Coal, issued in May 1994, includes only sixteen deep mines currently producing 27 million tonnes a year and with reserves of 450 million tonnes, together with seven 'care and maintenance' collieries no longer producing. The judgement is that

[13] *The Prospects for Coal*, pars 7.34, 7.40.

[14] Department of Trade and Industry, *The Energy Report: Markets in Transition*, London, DTI, 1994, para. 2.13.

[15] Parker, *The UK 'Coal Crisis'*, p. 45.

the remaining deep mines should be profitable during the remaining years of the contracts with the electricity generators (which continue to March 1998); these contracts provide for prices by the end of the period which represent a reasonable alignment with sustainable international steam coal prices, after taking account of the 'transport advantage' of most UK mines supplying inland power stations. There are a number of unresolved issues after privatization, including the uncertainties of new contracting arrangements after March 1998 and the financial robustness of the new pattern of ownership. However, in policy terms, the government could claim to have resolved the issue of uncompetitive output: after the sale of British Coal, these will be matters for the new owners, not for the government. No subsidies for private operators are contemplated.

Similarly, the rundown of manpower required to get the coal industry into a state where it could be sold to the private sector has already happened. Colliery manpower, which at the end of 1992/3 was 32,000, is now less than 10,000. The protection of jobs and the regional problems arising from *future* job losses are effectively no longer a policy issue for the present government. There is, of course, the question whether the issue would arise again in future if there were a change of government. While this cannot be ruled out, the political significance of coal mining has already been greatly reduced by the extent of the contraction to date: when the last Labour government was in office, there were 235,000 mineworkers, compared with fewer than 10,000 today. Moreover, increasing productivity and exhaustion of economic reserves (in the likely absence of major investment) will further reduce the mining labour force. In any event, it is difficult to envisage any change of UK policy which, by making use of state aids allowed under the new Decision, would materially affect the continuing decline of the already much reduced UK deep-mining industry.

Somewhat different considerations apply to opencast coal mining in the UK. In the past, costs of opencast coal have been well below those of deep-mined coal (see Table 2.3); UK opencast production had the lowest average cost of any in the EC, for the most part being competitive with internationally traded coal at inland power stations. Potentially, there are substantial reserves that can be produced profitably. The Preliminary Information Memorandum for privatization gives the reserves available at existing sites as 50 million tonnes – a reserve/production ratio of only four. Indeed, the average site has a life of only about four years, so that a quarter of all sites have to be renewed each year. The reserves available at prospective sites are put at about 200 million tonnes – but their exploitation depends upon obtaining planning permission on a site-by-site basis, usually against local environmental opposition, and often involving public inquiries. In the past, opencast mining has been seen as a source of cross-subsidy to sustain a large deep-mining industry, but this consideration will no longer apply after privatization. It may therefore be difficult to sustain UK opencast output in the range 10-15 million tonnes. In any case, protectionist measures to safeguard opencast production will not arise.

### 3.3 The future of protection: Germany

The German government's position is still based in large measure on the document agreed with the industry as part of the 1991 'coal round': Kohlekonzept 2005. This they would regard as representing the kind of restructuring plan called for by the Commission in the new Decision. The contrast with the UK government policy could hardly be more marked. Whereas the UK government has taken a view on the future of the international coal market (which is largely the same as that taken by the Commission), and has effectively completed a restructuring programme to make the UK coal industry competitive (at least within the UK), the German document does not even seek to address this issue: everything is submerged within the 'security of supply' argument, which is given no economic dimension. There is no discussion of how large the 'security premium' should be or whether it should be reduced. Instead, there is reliance on the idea of technical excellence: 'Today, hard coal-mining in Germany is an extremely modern, technologically advanced industry, marking it as a world leader. The high technology found in German mines today includes full automation, process control and computed assisted monitoring of operations.' As proof of this excellence the document states that, 'since the beginning of the 1980s, real costs in German hard coal-mining have remained constant.' There is no recognition of the very limited nature of an achievement which still left German coal costs at three times the level of international prices. Moreover, the reduction in output and manpower set out in

Kohlekonzept 2005 implies an increase in productivity of only about 20% to the end of the century – providing little if any erosion of the huge gap between German costs and international coal prices. The Germans therefore are implicitly admitting that the only way in which their hard coal industry can be sustained over the next decade (with 50 mtce of 'protected' production) is by subsidy, and therefore by political action. The question is, will the political support within Germany be eroded? The factors to be considered here are:

- whether the requirement made in the new state aids Decision that the *Kohlepfenning* be replaced by overt subsidy on the federal budget will weaken the industry's position, given the increasing pressures on the federal budget.
- whether the influence of North-Rhine/Westfalia and Saar land has been weakened following unification. The former GDR's lignite industry is being greatly reduced, and the disparity between its treatment and that of the highly protected hard coal industry in the former FRG is very evident.
- whether the institutional barriers to the large-scale expansion of gas into power generation in Germany will be eroded due to general trends towards more liberalized utility structures or pressure from industrial consumers for lower electricity prices.

Even before the new state aids Decision had been issued in December 1993, there were signs of some internal tensions in Germany which might call into question the robustness of Kohlekonzept 2005. By September 1993, it was reported that although Chancellor Kohl of the CDU officially stood by his agreement with Kohlekonzept 2005 of November 1991, the economic spokesman for the CDU/CSU coalition, Rainer Haungs, offered the view that open-ended subsidies for hard coal could no longer be afforded, and that the maximum amount of subsidy for steam coal should be decided, leaving the mining companies themselves to decide how much coal should be produced. In addition, the federal economics minister, Guenter Rexrodt (FDP) recommended increasing the *Kohlepfennig* levy on electricity prices from 7.5% to 8.5% in order to reduce the deficit on the *Kohlepfennig* fund; but this found no favour with his colleagues. Meanwhile, miners began protesting over the absence of agreement to an extension of the *Jahrhundertvertrag* after 1996, while some of the utilities indicated that they needed to plan ahead for imported coal supply.[16]

Agreement on these difficult issues was further hampered by the problems in arriving at a reciprocal agreement on the future of the nuclear industry, which is favoured by some non-coal producing *Länder*, particularly Bavaria, but opposed by the SPD and the coal-producing *Länder*.[17]

By November 1993 there had been further policy shifts by the German government. The *Kohlepfennig* levy rate was to be increased to 8.5% from 1994 and extended to 1996, after which the form of replacement taxation would have to be decided. But the issue of the future of the *Jahrhundertvertrag* contract still had to be decided. If it expired at the end of 1995 there would be a commitment to subsidy but no contractual arrangement to give effect to it.

These arrangements proved very unpopular with the Association of Electricity Utilities (VDEW), particularly because as the *International Coal Report* remarked, 'by fixing the level of subsidies but not the volume of coal production that these subsidies cover, the Federal government has shifted the responsibility of determining the level of future output from the producers to the electric utilities.'[18] In December 1993, the German government announced major changes which would take place when the *Jahrhundertvertrag* expires at the end of 1995: German coal producers would effectively no longer have a guaranteed market share, and power utilities be free to buy their coal from the most competitive supplier: however, German coal producers would be given a subsidy for steam coal to bridge the gap between prices and production costs of DM 7.5 billion in 1996, and DM 7 billion a year from 1997 to 2000.

The above developments are indicative of the underlying pressures on the German industry. Indeed, 1993 showed some contraction (see Table 3.1). The political uncertainties continued into 1994. In February 1994 the German parliament's upper house economics committee rejected the government's draft law as well as the further development of nuclear power. The committee decided that a subcommittee should examine another draft law drawn up by the coal-producing *Länder*, proposing that the guaranteed tonnages of steam coal included in Kohlekonzept 2005 be reinstated, with the public funds

---

[16] *International Coal Report* (ICR), No. 332, 20/9/93

[17] ICR No. 335, 29/10/93.

[18] ICR No. 337, 29/11/93

**Table 3.1 The German coal industry: summary, 1992-1993 (mtce)**

| | 1992 | 1993 |
|---|---|---|
| Production | 66.5 | 58.7 |
| Sales | 63.5 | 58.8 |
| Employees | 115,000 | 106,000 |

Source: International Coal Report No. 339, 10/1/94.

necessary to subsidize these tonnages being raised by an energy tax introduced in 1996 after the phasing out of the *Kohlepfennig*.[19] Nevertheless, in April/May 1994 the parliament passed the law committing the payment of steam coal subsidies of DM 7.5 billion in 1996, and DM 7 billion a year from 1997 to 2000. By means of this law, the federal government has moved from a volume to a funding structure, to provide some incentive to increases in efficiency in the mining industry, particularly as the subsidy limits are in nominal terms, so that the industry will have to absorb the effects of (modest) general inflation. It is still not clear how the steam coal subsidy is to be raised after 1996, or in what way the *Jahrhundertvertrag* is to be extended after 1995, given the reluctance of some utilities to enter into further long-term contracts. Further, in September 1994, the federal economics minister proposed that the federal government would in future pay only 50% of coking coal subsidies, instead of the two-thirds it has paid until now, with the balance falling on the coal-producing *Länder*.[20]

It was noted in Chapter 1 that the federal nature of the German constitution has been instrumental in defending the position of the German coal industry. Here, the powers and composition of the Bundesrat (federal council) are crucial. The Bundesrat has to be consulted on all legislation and has effective powers of veto on any measure which depends on the *Länder* for its implementation, and for that reason is regarded as affecting their rights and vital interests. The future of the coal industry would be so regarded by North-Rhine Westfalia (the most populous of all the *Länder*) and Saarland, and would be supported by other *Länder* controlled by the SPD. There are clearly tensions here, since

[19] ICR No. 342, 21/2/94
[20] ICR No. 357, 16/9/94

there is substantial cross-subsidy to the coal-producing *Länder* from the others, some of which, particularly Bavaria, are in favour of nuclear power. Moreover, the continuing economic and political pressures in the coal industry have led to increasing scepticism in political circles as to whether in fact a target of 50 mtce of 'protected' production is still realistic in the longer term. Nevertheless, given the continuing control of the Bundesrat by the SPD and its allies (who favour coal) and the narrowness of the victory of the CDU and its allies in the 1994 federal elections to the Bundestag (lower house), it seems unlikely that the position of the coal interest will be so weakened as to lead to an abandonment of the gradualist, consensus-seeking approach to coal policy adopted hitherto – although the 'price' might be reluctant SPD support for the continuation of nuclear power in Germany. The reunification process has had very little effect on this except to increase the budgetary pressure for coal subsidies to stay within the (very generous) limits proposed.

### 3.4 The future of protection: Spain

Table 2.3 above shows that Spanish deep-mined output is by far the most expensive category of hard coal production in the EU-12, with costs in 1992 over four times the commercial value relative to international coal prices. Most of the underground hard coal production comes from the Asturias/Leon region in northern Spain. The geology in this region is generally very difficult, the inherent ash contents very high (30-40%) and the calorific value low (averaging about 21 GJ/tonne). However, even within these averages, the position in the central Asturias coalfield is the worst. Between 1967 and 1970 the imminent failure of this industry led the government to take over the 18 private companies to form the state-owned Hulleros del Norte (HUNOSA). These are small-scale operations; in 1990 23 mines produced only 3.1 million tonnes of poor-quality coal – but they employed about 18,000 men. Productivity here was even lower than the Spanish average. The political pressure to support the Asturias region is of long standing, and coalfield representatives have been prominent in coal politics in Brussels, particularly in the European Parliament. Spanish coal policy is thus bound up with policy towards the Asturias region.

HUNOSA is probably the most uneconomic coal mining operation in the world. Even with prices twice the world level, 1991 losses were some $200 per tonne.[21] However, regional unemployment is high; and plans announced in late 1991 to close 9 out of the 23 pits and shed a third of the 18,000 workforce by the end of 1993 caused considerable opposition, so that a more gradual phasing was agreed. Nevertheless, between 1990 and 1993 six mines were closed and the workforce reduced by 3,000. For the foreseeable future, policy will be driven by the balance of pressures to reduce the enormous cost of coal production in the Asturias, and the continuing very difficult social/ regional problems involved in phasing out production.

### 3.5 The future of protection: France

Although there may be some social problems in Lorraine, there are no major political issues involved in the gradual phasing out of coal production in France. A continuation of present measures, within the framework of the new EU-12 state aids Decision, seems the most likely outcome. Charbonnages de France has signed a contract with Electricite de France (EdF) for the sale of 2 million tonnes per annum from 1994 to 1998, at prices linked to those of imported coal. The objective is the closure of all unprofitable mines by 2005, as is reflected in an agreement between the government and EdF.

---

[21] M.C.Daniel and E.Jamieson, *Coal Production Prospects in the European Community*, Paris: IEA Coal Research, 1992.

# Chapter 4

## The Environment and the Politics of Coal

In looking at the impact of environmental issues on the politics of coal in western Europe, it is essential to make the distinction between 'conventional' air pollution issues such as 'acid rain' and the global warming/climate change issue, since these are utterly different in the scale of difficulty involved in devising appropriate public policies, particularly at international level.

### 4.1 'Acid rain'

Chapter 1 has summarized the agreements and other measures taken by EU-12 governments to deal with the problem of 'acid rain' (in particular concerning $SO_2$ emissions), which ensure a significant overall reduction in emissions, and the special stringency applied to the emission standards required for new coal-fired plant. Nevertheless, they show significant variation, and will in practice require very few 'hard choices' by governments in the attainment of targets in reductions in $SO_2$ emissions. In particular, none of the LCPD targets for 2003 have required (or will require in the future) a restraint in the use of coal greater than individual governments would have wished, since, compared with the 1980 emission levels on which the targets are based:

- for France and Belgium a 70% reduction presents no problems, given the large increase in the nuclear component, and is unlikely to have any impact on policy;
- for the Netherlands, a 70% reduction presents little difficulty, given the move out of high-sulphur fuel oil into low-sulphur coal imports which has already occurred, the requirement that FGD be fitted to both existing and new coal stations, and the availability of natural gas for power generation;
- for Italy, a 63% reduction is unlikely to have an impact on the use of imported coal for power generation (sulphur limits on the coal are already

in place), particularly in view of the prospect of increased use of gas and low-sulphur fuel oil;

- for Denmark, a 67% reduction is unlikely to affect coal policy, given the large-scale move out of high-sulphur fuel oil into low-sulphur coal imports which has already occurred, and the planned increase in the use of natural gas, which (in Denmark's case) is probably desirable in order to reduce the unusually high dependence on coal in the generating system;
- for Spain, there is a much lower target reduction of 37% which, given the reduced use of high-sulphur fuel oil and the increased nuclear component which have already occurred, is unlikely to restrict the use of coal to a degree which would endanger government policy on Spanish coal production, even after allowing for the high rate of growth in Spanish electricity production;
- for Greece, Ireland and Portugal, which are 'allowed' *increases* of 6%, 25% and 79% respectively, there are no significant implications for the use of hard coal.

On 'acid rain' issues, the most revealing contrast is between the UK and Germany, where the $SO_2$ emission reduction targets for existing plant are 60% and 70% respectively by 2003, compared with 1980, with a slower interim rate of reduction for the UK. In the UK, where, at the time the LCPD was agreed, the electricity generating system was mainly dependent upon relatively high-sulphur (average 1.6%) indigenous coal, only a modest FGD programme of 6 GW is being put in place, whereas in Germany, where the sulphur content of indigenous coal was lower, all coal-fired power stations were fitted with FGD. In effect, the 'acid rain' issue has been removed as a threat to the use of coal in Germany, because with FGD fitted to all power stations, the sulphur content of the coal becomes of far less significance. However, in the UK, $SO_2$ is still an issue. This is not because the target of 60% is in jeopardy, since such a target is consistent both with the rapid rundown of the coal industry which has already taken place and with the 'dash for gas', both of which, as we have seen, were the direct results of government policy. Rather, the issue concerns the operation of BATNEEC as applied to existing coal-fired power stations. When Her Majesty's Inspectorate of Pollution (HMIP) gave generating plant authorizations on emission limits in 1993, they required that the two major generators should put forward proposals in 1994 which would enable each power station to achieve the more stringent emission requirements for new plant as soon as possible, and in any event not later than April 2001.

If this policy guideline were taken literally, it would mean effectively that all continuing coal-fired plant would have to be fitted with FGD, or replaced. Yet, under present plans, only 6 GW out of the 23 GW of coal-fired plant likely still to be in operation in the late 1990s will be fitted with FGD, and most of the balance will be over 30 years old by the end of the century. Given the high cost of FGD retrofits, such investment is very unlikely to be made by the now privately owned generators. Moreover, HMIP has expressed a preference for gas rather than FGD as representing BAT. Such a situation would most likely lead to a second 'dash for gas'. However, the application of the BATNEEC principle in this case is not clear-cut. The 'NEEC' part of the formula is intended primarily as a means of providing *reasonable time* to meet 'BAT', not as a means of avoiding it; but in practice it might prove possible for some of the existing coal-fired power stations to continue, provided that they operated at low load-factors, and that national objectives (and the 'emission bubbles' allowed to main generators) were met, probably entailing increased use of low-sulphur coal imports. The formulation on the BATNEEC issue in the UNECE Second Sulphur Protocol suggests that such a compromise is likely. The introduction of tradeable permits would facilitate such an approach. But the important point for this study is to draw attention to the contrast between the comprehensive forward protection from 'acid rain' regulation afforded to the German coal industry through the universal FGD policy, and the continuing uncertainty attached even to the greatly reduced use of UK power station coal by reason of $SO_2$ regulations (particularly if these regulations are tightened). The UK targets in the UNECE protocol, requiring an 80% reduction by 2010, will maintain this pressure. The reasons for this contrast are to be found (once again) in the different coal politics of Germany and the UK.

The timing of the LCPD, and the establishment of $SO_2$ reduction targets throughout the EU-12, was in large measure due to the influence of (West) German politics. As described elsewhere, in 1981/2 there was intense media interest in the 'threat to the German forests' from air pollution, and this coincided with electoral successes of the Green Party at regional and European levels which threatened the survival of the small FDP (Free Democratic Party), the 'kingmaker' of German politics within the CDU/CSU/FDP coalition, which was confirmed as the government in March 1983.[22] The electorate wanted more effective environmental protection and demanded immediate action to secure the forests. Inherited FDP draft legislation to clean up coal-fired power stations thus became a centrepiece of the new government's 'Save our Forests' campaign. This quickly became law in July 1983 and was taken up very shortly afterwards by the Commission of the European Communities, which in early 1984 issued a draft Directive closely resembling German legislation. By 1986 the Federal Emission Control had been further strengthened to enable regional authorities to apply strict emission standards to virtually the entire plant stock. Effectively all the German political parties had competed to show that they were taking the most effective action against acid rain.[23] The action also replaced the uneasy political balance between the pro-coal SPD and its allies and the pro-nuclear CDU and its allies – a polarity expressed in *Land* terms by North-Rhine/Westfalia and Bavaria.

In this way, action within (West) Germany which was designed to achieve the double objective of being seen to be taking rapid and effective action to 'save the German forests', while at the same time protecting the German hard coal industry from the consequences, was turned into Community-wide pressure for other countries to take similar action (thereby minimizing the cost disadvantage incurred by (West) Germany). At the time the LCPD was agreed in 1988, it was generally assumed that to sustain UK coal production would require 12 GW of retrofitted FGD on coal-fired stations. However, as electricity privatization approached, it became clear that 12 GW of FGD was not a firm commitment. The House of Commons Energy Select Committee investigated the position in 1989-90 and reported that the views of the European Commission given in evidence 'forces us to the conclusion that the Government obtained relatively undemanding limits for the UK on the understanding that the UK would continue the required reductions [in $SO_2$ emissions] chiefly through FGD, and that having obtained such limits by that means the UK

---

[22] Sonja Boehmer-Christiansen, *The Politics of the Environment and Acid Rain in the Federal Republic of Germany: Forests versus Fossil Fuels?* University of Sussex: Science Policy Research Unit, 1989.

[23] Ibid., pp. 151-2.

now proposes to comply with them by cheaper methods instead'.[24] The 'cheaper methods' were gas-fired CCGTs and imported low-sulphur coal. The UK government's emphasis was on market forces. 'The long-term challenge for British Coal is to make UK coal the fuel of choice, and FGD the method of sulphur abatement which the generators would wish to choose on economic grounds.'[25] As we have noted above, this was a forlorn hope. At present only 6 GW of FGD is installed or under construction, giving environmental security to only 12-15 million tonnes of UK coal (compared with the 35 mtce under the forward arrangements in Germany). In this matter, the UK government's main concern was to require as little FGD as seemed compatible with obligations under the LCPD, and thereby minimize the financial liability to be placed on the privatized generators.

### 4.2 Clean coal technology

An alternative to FGD is the development of advanced/'clean coal' generating plant. Yet the degree to which such development will mitigate the 'acid rain' problem – at least for some time ahead – is doubtful. Evidence put forward at the time of the 'UK coal crisis' indicated that, because of the substantially higher capital costs of coal-fired plant as compared with CCGTs, gas prices would have to rise some 50% above 1993 levels, even with constant coal prices, for advanced clean coal plant to be the preferred option. In the event of carbon taxes or other measures to restrain $CO_2$ use being introduced, the gas price increase would have to be even greater. While these figures may not be wholly appropriate for all EU-12 countries (because of differences in the structure and costs of their gas industries), nevertheless advanced clean coal plant will face significant economic disadvantages unless or until the gas price for power generation rises substantially relative to coal.

We have already noted that the totality of coal-fired generating plant falls far short of the amount required to stabilize the total market for coal into the next century. Indeed, for coal the main long-term issue will be the extent to which new coal-fired generating capacity can be made compatible with

[24] House of Commons Energy Select Committee, *The Flue Gas Desulphurisation Programme* (HC371), London: HMSO 1990, para. 52.

[25] Ibid., Memoranda of Evidence, p. 23 para. 21-3.

environmental concerns. If clean coal technology were to be the answer, then something like 20 GW of 'clean coal' plant would now need to be at various stages of construction. In fact, the figure is less than 1 GW.[26]

The lack of a sense of urgency on the part of governments in expediting the introduction of clean coal technology arises from a combination of factors:

- The perceived abundance of gas, and the availability of commercially proved CCGT technology, have made this the obvious choice for economic and environmentally friendly generating plant. By contrast, coal combined-cycle plant is still to be fully proved commercially. The capital cost is much higher than for gas, and economic operation requires a large increase in the price of gas which currently does not seem likely.
- Privatization and liberalization in the electricity market have increased the difficulty for private commercially oriented companies to provide large funds for R&D and development for which they can obtain little if any direct competitive advantage in the market; and for coal producers, new technology will benefit imports as much as indigenous production. This places an onus on government or the EU to provide funds: but they have generally been unwilling to provide funds which are not matched by the private sector.
- Only Germany has a comprehensive policy of protecting its coal industry, yet in Germany the industry is protected for some time ahead by the policy of universal FGD and other emission control measures already taken. This has removed any urgency for clean coal technologies, even in Germany.

It may be objected that in fact substantial sums of public money have been spent on clean coal technology over the years – not least by the European Community. In a document published in 1992, *Coal can be Green*, the Commission (DGXVII) make the following points:

(1) The EC provided substantial financial support to assist pilot and demonstration plants in the period before 1989, in the following amounts (million ECU):

---

[26] IEA, *Coal Information 1993*, Paris, 1994.

| | |
|---|---|
| Fluidized bed combustion | 39 |
| Coal combined cycles | 27 |
| Coal gasification | 123 |
| Coal liquefaction | 41 |
| Total | 230 |

These amounts generally made up 40% of the total eligible cost of the programmes supported.

(2) The THERMIE programme set up to cover the period 1990-4, with a total budget of 700 million ECU, has a strong coal element, with the EU providing up to 40% of the eligible costs in 'innovative' projects and 35% in 'dissemination' projects. The hard coal projects supported included:

- the liquid solvent extraction process of coal liquefaction at Point of Ayr (UK);
- circulating fluidized bed boilers in France and Spain;
- the British Gas/Lurgi slagging gasifier at Westfield (Scotland);
- the Krupp Koppers PRENFLO gasifier at Furstenhausen (Germany);
- integrated gasification combined cycle plant (IGCC) – the decision in 1991 to support a demonstration plant in Spain at Puertollano as a European project supported also by Spanish, French and Portuguese utilities, and testing a wide range of EU-12 and imported coal, at a total cost of 565 million ECU.

These programmes, financed by the Community, member states and enterprises both public and private, have nevertheless not yet achieved the breakthrough necessary to bring about a large programme of new advanced technology coal stations. The overall picture is still one of a large body of conventional generating plant, some of which has been retrofitted with FGD. It is unlikely that all existing plants will be brought up to the stringent standards required for new plant, or that they will be replaced (except at the margin) by 'clean coal technology' – at least not for the foreseeable future. This could have major implications in the longer term as the existing stock of power stations approach the end of their useful lives – either coal use (whether of indigenous coal or imports) will decline, or environmental standards will have to be relaxed.

### 4.3 Climate change and carbon dioxide

Since emerging as a major political issue in the later 1980s, concerns about climate change have represented a pervasive baleful influence on the prospects for the most carbon-intensive fuel, coal, particularly in western Europe. In 1992 the European Community and all other western European countries signed the Framework Convention on Climate Change, which has as its objective 'stabilization of greenhouse gas concentrations in the atmosphere at a level that would prevent dangerous anthropogenic interference with the climate system'. In May 1992, the European Commission adopted a proposal on a Community strategy to reduce $CO_2$ emissions, embodying an earlier decision to stabilize $CO_2$ emissions in the EU in 2000 at their 1990 levels, by means of a package of measures including modification in the energy sources used. The target is for the EU as a whole and would not necessarily have to be met by all individual member states.

At the time, the Commission expected that a 'business-as-usual' scenario would result in an 11% increase in $CO_2$ emissions in the EU by 2000, about half of which might be avoided by new initiatives on energy conservation (SAVE), renewable energy (ALTENER) and other new energy technologies (THERMIE). The balance could, in the view of the Commission, best be covered by national programmes, and an EU $CO_2$/energy tax. The logic here was that the 50% energy component of the proposed tax would, by its impact on energy prices, stimulate energy efficiency generally; while the 50% $CO_2$ component of the tax would stimulate the switch out of carbon-intensive fuels – the level of the tax being equivalent to \$3 per barrel of oil, rising to \$10 per barrel over time. This might result in saving 4% of $CO_2$ emissions by 2000. The 50:50 split between 'energy' and '$CO_2$' was the product of a compromise, basically between French and German interests: there was anxiety that a tax based solely on $CO_2$ would provide too obvious an advantage to France (with its overwhelmingly nuclear-based electricity system) as compared to Germany (with its desire to support its coal industry).

But the politics of coal were never far below the surface. In its document, *Coal can be Green*, DGXVII pointed out that coal-fired power stations worldwide provide only about 8% of the total 'potential' greenhouse effect. Nevertheless, in the run up to the Rio conference, much of the debate at Community level seemed (at least to this observer) to be based on the overriding need to reduce coal use in power stations.

The formulation of specific proposals for a carbon/energy tax generated more measured analysis, best summarized in the Commission's October 1992 document on the *Consequences of the Proposed Carbon/Energy Tax* (DGXVII (SEC (92) 1996) 1992. Detailed figures of $CO_2$ emissions, by fuel and sector, enable a more balanced picture to be drawn. Thus, it is clear that in 1990, only about a third of total $CO_2$ emissions in the EU-12 came from solid fuel, and only about a quarter from hard coal (after deducting the contribution from lignite and brown coal). Hard coal power stations account for less than 20% of total $CO_2$ emissions. There is however considerable variation between countries. In 1990, the proportion of $CO_2$ emissions attributed to hard coal varied from 8% in Greece to 45% or more in Denmark and Ireland (Table 4.1).

There is no correlation between the 'energy $CO_2$ intensity' of the various economies and the scale of emissions from hard coal, as can be seen by comparing Tables 4.1 and 4.2. Nor is there any correlation between these figures, and the $CO_2$ reduction targets adopted by individual member states. In other words, the present level of hard coal use has not been a decisive factor in determining the stance taken on the determination of targets. Other factors weigh at least as heavily. In the Commission's 'reference case' in SEC(92) 1996, between 1990 and 2000, in nearly all member states, the $CO_2$ emissions from transport oil are expected to increase substantially more than those from solid fuels (coal and lignite) – indeed, whereas the total $CO_2$ emissions from solid fuels are projected to decrease by 5%, those from transport oil are projected to increase by 25%. Given that the politics of restraining the growth of transport fuel use are much more difficult than those of fuel-switching in power station feed-stocks, this may in time throw the burden back on to coal; but in the meantime, there is little evidence that coal use in member states is being restrained by $CO_2$ targets, although these may be

**Table 4.1 $CO_2$ emissions and hard coal intensity, 1990**

| | Total $CO_2$ emissions | % derived from hard coal |
|---|---|---|
| Ireland | 31 | 46 |
| Denmark | 53 | 45 |
| UK | 579 | 41 |
| Belgium | 112 | 29 |
| Spain | 211 | 27 |
| Luxembourg | 13 | 26 |
| Germany | 1005 | 25 |
| Portugal | 40 | 25 |
| Netherlands | 157 | 19 |
| France | 366 | 18 |
| Italy | 402 | 12 |
| Greece | 74 | 8 |

Source: Author.

quoted in support of changes which were being undertaken for other reasons. The most important example here is the UK 'dash for gas'.

However, the compatibility of $CO_2$ targets for the year 2000 with projected patterns of hard coal production and use is largely a matter of the use of existing (or committed) capital stock. The real change in policy to the detriment of coal arising from $CO_2$ targets is likely to occur in the period after 2000, as a result of possible effects on the pattern of power generation investment. As the Commission's paper on the carbon/energy tax proposals, says 'While the energy imports resulting from taxation might indeed be limited up to 2000, nevertheless the longer term implications are likely to be of importance with a marked preference being expressed by electricity producers away from higher carbon intensive fuels.'

Attention has already been drawn to the fall-off in all coal-fired generation plant renewal, and within that the very small progress being made by clean coal technology. Even clean coal technology produces more $CO_2$ than CCGTs, so that any imposition of carbon-related taxes would worsen the relative position of coal-fired plant, thus adding to coal's higher capital cost disadvantage. Thus the building of new coal-fired plant to replace the large stock of obsolescent pulverized fuel coal-fired plant (even if fitted with FGD)

**Table 4.2 Energy-related $CO_2$ emissions per unit of GDP ($CO_2$/1000t ECU)**

| | Actual 1990 | Reference case 2000 |
|---|---|---|
| Luxembourg | 1.82 | 1.49 |
| Greece | 1.42 | 1.46 |
| Ireland | 0.92 | 0.83 |
| Portugal | 0.85 | 0.82 |
| Germany | 0.79 | 0.69 |
| UK | 0.75 | 0.66 |
| Belgium | 0.74 | 0.64 |
| Netherlands | 0.72 | 0.63 |
| Spain | 0.54 | 0.50 |
| Denmark | 0.57 | 0.50 |
| Italy | 0.47 | 0.42 |
| France | 0.39 | 0.37 |
| EU-12 average | 0.63 | 0.55 |

Source: Author.

would require not only the full commercialization of clean coal technology but also a large widening of the thermal price differential between coal and natural gas. *It is difficult to see any of the member states making policy interventions to ease the position of hard coal relative to natural gas in power generation investment, unless there is a dramatic and sustained 'security of supply' shock in respect of Russian/Algerian gas.*

In theory, while the loser from concerns about the climate change is hard coal (and lignite), there should be a clear gain for nuclear power, which in turn would reinforce the downside for coal. But is this likely? This is not the place to replicate the excellent recent study by Peter Beck for the RIIA, but three statements can be made without too great a fear of contradiction.[27]

- Those countries in western Europe which do not have operating nuclear capacity (Austria, Denmark, Greece, Ireland, Luxembourg, Portugal,

[27] *Prospects and Strategies for Nuclear Power*, London: RIIA, 1994.

Norway and now Italy) will not embark upon a programme of new nuclear stations for the foreseeable future. 'Coal versus nuclear' is therefore not an issue in these eight countries.

- With the exception of France, those countries with existing nuclear capability (UK, Belgium, Germany, Spain, Switzerland, Sweden, Finland and the Netherlands), seem most unlikely to start building additional new nuclear stations before 2000.
- The greater the extent to which governments follow the line taken by the UK government in privatizing electricity generation and subjecting it to competitive pressures, and in particular commercial rates of return, the lower will be the probability of new nuclear stations being built.

Thus, although there has been much discussion of the benefits to the nuclear industry of current generally held views on the need to cut greenhouse gas emissions (particularly $CO_2$), the above considerations make it unlikely that there will be major changes in public policy on nuclear power in any of the western European countries, at least for some time to come. A very heavy carbon tax could, of course, alter the relative economics of nuclear energy against coal, but the experience with the abortive Commission proposals for a diluted energy/carbon tax shows the political difficulty of introducing such measure.

In the four coal-producing countries, the climate change issue is unlikely to have any effect on the balance between coal and nuclear power *compared with the situation which would have arisen had there been no concern about $CO_2$ emissions, for the following reasons.*

- In the UK, although in the past the government sought to promote nuclear power, at least in part as a means of reducing the power of the coal-mining unions, given what has happened to the coal industry, this no longer applies. The terms of reference of the *Nuclear Review* (issued in May 1994) state that 'the government believes that the future of role of nuclear power in the UK's electricity supply will depend on it proving itself competitive while maintaining rigorous standards of safety and environmental protection'; and in the case for new nuclear stations, a 'key test' will be whether such stations can be built with private sector finance. Moreover, in the UK government's report *Climate Change: The UK programme*[28] there is studied avoidance of a commitment to additional nuclear power stations.

- In Germany, the problem of internal political balance between pro-coal and anti-coal *Länder* and parties will continue to be the crucial issue. The $CO_2$ emission reduction targets are unlikely to affect this. Rationalizations of the position whereby the German government follows a policy of both heavily subsidizing coal and adopting $CO_2$ emission reduction targets are readily available. For example, Kohlekonzept 2005 states blandly: 'In environmental terms: Kohlekonzept 2005 entails a 30% reduction of $CO_2$ emissions from German hard coal by the year 2000 in comparison with 1987. Thus German hard coal more than meets political $CO_2$ reduction goals (minus 25% by 2005) at an early date.'
- In France, the marginalization of coal use in a generation system dominated by nuclear power and the planned phasing out of domestic coal production by about 2005 are policies independent of $CO_2$ reduction targets.
- In Spain, the problems of public acceptability for the existing nuclear stations will not be affected either by concern for the Spanish coal industry or by $CO_2$ targets.

The real environmental threat to coal over the next 10/15 years will come not from nuclear power but from natural gas, where economics currently appear to go hand in hand with environmental protection, and where there is no problem of public acceptability.

---

[28] *Climate Change: The United Kingdom's Report under the Framework Convention on Climate Change*, London, HMSO, 1994.

# Chapter 5

## Policy Synthesis: Prospects and the Politics of Coal in Europe

I noted in Chapter 1 above that 'conventional' projections, based on unchanged policies, suggest some recovery in the western European market for hard coal by 2005, compared with 1993. However, more recent developments, particularly in the UK (where the inroads of gas into power generation might prove to be greater), in Germany (where the coking coal market is shrinking) and in Italy (where new coal-fired capacity has become speculative) could mean that the total market is unlikely to be above the bottom of the range quoted – that is, no more than in 1993 – and could well be less, particularly if CCGTs begin to make a major impact on the power generation market. Moreover, the 'conventional' projections, which show EU-12 hard coal production declining by a third between 1993 and 2005, with two-thirds of the western European coal market met by imports, could still understate the extent of the decline in indigenous production.

A centrally directed EU-12 policy to protect indigenous production has never been a serious possibility, because of the lack of common interest in such a policy. Coal producer states, which are in a minority, have even less influence on this issue with the enlargement of the Union; and the further enlargement currently proposed (to include Finland, Sweden, Austria and Norway) involves states with no significant coal production. (Indeed, the only enlargement issue which might have an effect on EU coal production policy would be the accession of Poland, whose hard coal output by then would probably exceed the total EU-12 output.) Most member states are concerned to ensure that they have free access to (generally) cheap imported coal from countries outside the EU-12. Moreover, because the cost of nearly all EU-12 production has been higher than the price of imports, the amount of intra-EU trade in EU-produced coal is minimal, so there are no significant Single Market implications. This has given rise to a situation where the 'subsidiarity principle' drives the policy of member states on coal production.

However, as we have seen, the results have been very different. The UK has gone for rapid contraction and privatization, France for slow phasing out, Germany for massive subsidization and slow adaptation by consensus and Spain for selective restructuring and subsidization. Moreover, the policies of individual governments have had no practical effect on the policies of others. There is no sense in which the German policy of high subsidies has contributed to the rapid decline of the UK coal industry; and it appears very unlikely that in practice what has happened to the UK industry will have any effect on the German industry. These matters have been, and will continue to be, determined solely by domestic policies.

What I have called the 'Anglo-German anomaly' continues to be particularly striking. In addition to the support for German coking coal, the subsidy for steam coal will be more than twice the price available to the UK coal industry; yet the broad aim of German policy is to sustain a level of protected deep-mined production twice that likely in the UK, even though German costs are likely to remain some three times the UK level. This huge difference stems from very sharp contrasts in the political influences which have brought it about, including in particular the following:

- the political dimension to the UK Conservative government's attitude to the National Union of Mineworkers, and the extreme confrontational stance taken by the miners' leader, Arthur Scargill. By contrast, the German mining union has been a model of moderation, and a willing participant in the joint consultation and consensus approach adopted within the German coal industry, and in its relationship with the Federal government.
- the checks and balances in the federal structure of German politics, which sustained the influence of the SPD-dominated coal-producing *Länder*. The UK, with its highly centralised policy-making process, had no such regional checks of sufficient influence to mitigate the rigour of central government policy.

Now that the threat of OPEC appears to have receded, and future gas supplies appear abundant, the political influences in favour of coal use (as distinct from those in support of indigenous production) have virtually disappeared, and have been replaced by environmental concerns relating to coal combustion.

Tony Brenton, in his recent book *The Greening of Machiavelli* has given a number of reasons why, on environmental policy, member states of the EU-12 'find themselves under a number of pressures to adjust their national priorities to respond to the concerns of their Community partners'.[29] Thus, on general grounds it might have been expected that environmental policy towards coal in the EU-12 would have a stronger element of collective action than had policy on coal production. Yet, in practice, we have seen that, although there has been a significant EU element, notably in the Large Combustion Plant Directive, there is no evidence that any environmental regulation has been adopted which would limit any member state's ability to use coal (or indeed any other fuel) against its government's wishes. On the other hand, the absence of a centrally directed EU-12 energy policy, and the current lack of support for 'strategic' arguments in favour of coal use, have made it inconceivable that the Commission would be able (or indeed wish) to intervene in the market to shield coal from its actual or potential environmental disadvantages. It is also highly improbable that any Member State would intervene in this way, except in support of a policy linked to a political imperative to protect indigenous production. Environmental policy towards coal use has been marked by the subsidiarity principle in both coal-producing and other member states.

Might this change in future? We have seen how the picture over the next decade is largely determined by existing capital stock and decisions already made. For the longer term, a discussion paper prepared by the Commission in 1993 quotes two external studies that show the potential impact of a 'strong environmental policy scenario' on coal's share of European primary energy by 2015/2020 (see Table 5.1).[30] As the Commission points out, where energy futures are driven by strong environmental policies there are substantial changes in fuel mix, radically reducing coal and oil use. Increasing environmental concerns could limit coal use, whereas use of gas might be limited because of energy security constraints and that of nuclear power by safety and waste disposal concerns.

Subsequently (April 1994), DGXVII held a workshop on 'Energy Futures: The Economic Foundations', including a discussion on energy sources and

---

[29] Tony Brenton, *The Greening of Machiavelli*, London, RIIA/Earthscan, 1994, p. 110.

[30] European Commission Directorate-General for Energy (DGXVII), *New Orientations in Energy: The Economic Foundations*, Brussels, September 1993.

**Table 5.1 Coal's share of European primary energy by 2015/2020 under different scenarios (%)**

| | 2015[a] | 2020[b] |
|---|---|---|
| 'Strong environmental policy' scenario | 8 | 11 |
| All other scenarios (average)[c] | 18 | 20 |

[a] *Scanning the Future*, Dutch Planning Bureau, 1992;
[b] *Energy for Tomorrow's World*, World Energy Council, 1993;
[c] (Share in 1990 = 22%).
Note: these numbers include brown coal.

geopolitics. The present author made a presentation on coal, against the background that in ten years' time the EU could be importing some three-quarters of its coal requirement. Nevertheless, I argued that there was little geopolitical risk on imported coal supply, as:

- world coal reserves were abundant and widely distributed, and western European imports represented only about 7% of world hard coal output;
- there was no possibility of a 'COALPEC' or cartelized pricing; coal trade had a highly competitive structure with prices determined largely by coal-on-coal competition, with trade mainly a matter for companies rather than governments;
- since the second OPEC price shock of 1979, there had been a five-fold increase in seaborne steam coal trade, but prices had fallen in absolute terms (and substantially in 'real' terms); this was a commodity trade which tended to over-supply and operated on small profit margins;
- problems in particular countries (e.g. South Africa) would be likely to be readily offset by additional supplies from elsewhere (particularly the US).

Indeed, the geopolitical risk on coal was less than for other fossil fuels (or nuclear power). In these circumstances, in weighing the balance between environmental and economic considerations, the EU-12 would need to think very carefully before it abandoned the low strategic risks of coal.

Whether such weight will be given will be determined by political considerations, either at EU level or at member state level. The environmental disadvantages of coal will be important if the political pressures to take action on climate change outweigh other factors. Put thus, the point is very obvious; but much of the discussion on the future of coal (and, indeed, nuclear power) has been conducted as though there were a readily quantifiable 'cost/benefit' approach which could determine an optimum balance between the conventional economic calculus and environmental costs. This is not the case, and will not be for the foreseeable future: so these matters will be determined by politics. We have shown how diverse these political considerations can be in the case of hard coal in western Europe. Tony Brenton has reminded us of the unpredictability of public opinion on these matters. He says:

> 'Modern sociology offers the model of the 'issue/attention cycle' by which a public issue, such as ecology, for some reason captures popular attention and then becomes the object of a crescendo of public alarm as the concern it provokes becomes self-generating, both through simple osmosis and by drawing to public attention more and more instances of the problem under discussion. Eventually, however, the bubble deflates as the cost and difficulties of action become apparent and public interest moves elsewhere, often leaving the original problem unresolved, and the public as indifferent to it, as before the cycle began'.[31]

This, of course, is not a universal law of nature, but is one possible pattern of the influences brought to bear on democratic governments. Further, on those issues which are international, Brenton makes a particularly apposite quotation from de Tocqueville: 'As a rule democracies have very confused or erroneous ideas on external affairs; and generally solve outside questions only for internal reasons.'[32] Experience to date strongly suggests that the striking of the balance between environmental and economic factors in coal use will continue to have to be compatible with the interests of individual member states, however *communautaire* the form of policy may be.

This study has also served to illustrate how the balance of opposing internal political forces determines the rate of change – with particular emphasis on

---

[31] Brenton, *The Greening of Machiavelli*, p. 24.
[32] Ibid., p. 110

the contrast between coal production policy in the UK and in Germany. In the case of the UK, energy economics, the government's environmental policies and UK 'energy politics' all pressed in the same direction, so that once the government's fear of the NUM had been effectively removed by the defeat of the 1984-5 strike, rapid rationalization of the coal industry was achieved; in Germany, on the other hand, notwithstanding the much poorer economic position of the coal industry, political and institutional factors have acted as a very substantial brake on the rate of change.

If there were a serious nuclear accident, if alarm were to grow about the security of gas supplies from Russia and Algeria, and if climate change were no longer seen to be a major threat, then the politics of coal could be changed sufficiently to stimulate investment in new coal-fired generation capacity (including 'clean coal technology') in the first half of the twenty-first century. Such a scenario, by sustaining coal consumption, might also slow the rate of contraction in the remaining EU-12 hard coal production. But, making a judgement in 1994, the same conclusion can be reached of the EU-12 coal industry that I reached concerning the UK coal industry at the time of the 1992 'coal crisis': that all this is not about fundamental strategic direction, but about politically acceptable phasing of decline. As we have seen, this process has been governed by domestic politics operating within the subsidiarity principle, and, in the case of coal, has led to outcomes so different in the UK and Germany as to strain the bounds of credibility – until we remember the politics.